Studies in Computational Intelligence 439

Editor-in-Chief

Prof. Janusz Kacprzyk
Systems Research Institute
Polish Academy of Sciences
ul. Newelska 6
01-447 Warsaw
Poland
E-mail: kacprzyk@ibspan.waw.pl

T0181209

For further volumes:
http://www.springer.com/series/7092

Cristian Lai, Giovanni Semeraro,
and Eloisa Vargiu (Eds.)

New Challenges
in Distributed Information
Filtering and Retrieval

DART 2011: Revised and Invited Papers

 Springer

Editors

Cristian Lai
CRS4, Center of Advanced Studies
Research and Development in Sardinia
Parco Scientifico e Tecnologico
della Sardegna
Pula
Italy

Eloisa Vargiu
Department of Electrical and
Electronic Engineering
University of Cagliari
Cagliari
Italy

Giovanni Semeraro
Department of Informatics
University of Bari "Aldo Moro"
Bari
Italy

ISSN 1860-949X
ISBN 978-3-642-43857-8
DOI 10.1007/978-3-642-31546-6
Springer Heidelberg New York Dordrecht London

e-ISSN 1860-9503
ISBN 978-3-642-31546-6 (eBook)

Preface

Information filtering (IF) has drastically changed the way information seekers find relevant results. Information seekers effectively prune large information spaces and help users in selecting items that best meet their needs, interests, preferences and tastes. These systems rely strongly on the use of various machine learning tools and algorithms for learning how to rank items and predict user evaluation. Information Retrieval (IR), on the other hand, attempts to address similar filtering and ranking problems for pieces of information such as links, pages, and documents. IR systems generally focus on the development of global retrieval techniques, often neglecting individual user needs and preferences. This book focuses on new challenges in distributed Information Filtering and Retrieval. It collects invited chapters and research contributions from the DART 2011 Workshop, held in Palermo (Italy) and co-located with the *XII International Conference of the Italian Association on Artificial Intelligence*. DART aimed to investigate novel systems and tools to distributed scenarios and environments. Therefore, DART contributed to discuss and compare suitable novel solutions based on intelligent techniques and applied in real-world applications.

Chapter 1, *A Brief Account on the Recent Advances in the Use of Quantum Mechanics for Information Retrieval*, is an extended contribution by Massimo Melucci, invited speaker at DART. The chapter focuses on the attention and research that have been paid to the exploitation of Quantum Mechanics in IR. It introduces the basic notions of Quantum IR and explains how the retrieval decision problem can be formulated within a quantum probability framework in terms of vector subspaces rather than in terms of subsets as it is customary to state in classical probabilistic IR. Hence it shows that ranking by quantum probability of relevance in principle yields higher expected recall than ranking by classical probability at every level of expected fallout and when the parameters are estimated as accurately as possible on the basis of the available data.

In Chapter 2, *Cold Start Problem: a Lightweight Approach*, the SWAPTeam1 participation at the ECML/PKDD 2011 - Discovery Challenge for the task on the cold start problem focused on making recommendations for new video lectures.

The Challenge organizers encouraged solutions that can actually affect VideoLecture.net. The main contribution concerns about the compromise between recommendation accuracy and scalability performance of proposed approach. For facing the cold start problem, authors developed a solution that uses a content-based approach. Such an approach is less sensitive to the cold start problem that is commonly associated with pure collaborative filtering recommenders. The surrounding idea for the proposed solution is that providing recommendations about cold items remains a chancy task, thus a computational resource curtailment for such task is a reasonable strategy to control performance trade-off of a day-to-day running system.

Chapter 3, *Content-based Keywords Extraction and Automatic Advertisement Associations to Multimodal News Aggregations*, focuses on multimodal news aggregation retrieval and fusion. In particular, authors tackle two main issues: extracting relevant keywords to news and news aggregations, and automatically associating suitable advertisements to aggregated data. To achieve the first goal, authors propose a solution based on the adoption of extraction-based text summarization techniques; whereas to achieve the second goal, they developed a contextual advertising system that works on multimodal aggregated data. The proposed solutions have been assessed on Italian news aggregations and tested with suitable baseline solutions.

In Chapter 4, *ImageHunter: a novel tool for Relevance Feedback in Content Based Image Retrieval*, a Content-Based Image Retrieval (CBIR) engine, called ImageHunter, is thoroughly described. ImageHunter aims at providing users, especially unskilled ones, with an effective tool for image search, classification and retrieval within digital archives, photo-sharing web sites and social networks. The novelty of the approach, with respect to canonical techniques based on metadata associated to images, lies in the combination of content-based analysis with feedbacks from the users (Relevance Feedback). In addition, the modular structure permits the integration of ImageHunter in web-based applications as well as in standalone ones.

Chapter 5, *Temporal Characterization of the Requests to Wikipedia*, presents an empirical study about the temporal patterns characterizing the requests submitted by users to Wikipedia. The study is based on the analysis of the log lines registered by the Wikimedia Foundation Squid servers after having sent the appropriate content in response to users requests. The analysis has been conducted regarding the ten most visited editions of Wikipedia and has involved more than 14,000 million log lines corresponding to the traffic of the entire year 2009. The conducted methodology has mainly consisted in the parsing and filtering of users requests according to the study directives. As a result, relevant information fields have been finally stored in a database for persistence and further characterization.

In Chapter 6, *Interaction Mining: the new frontier of Customer Interaction Analytics*, authors present a solution for argumentative analysis of call center conversations. Their challenge is to provide useful insights for enhancing customer interaction analytics to a level that will enable more qualitative metrics and key performance indicators beyond the standard approach currently used. These metrics rely on understanding the dynamics of conversations by highlighting the way participants discuss about topics. In so doing, relevant situations (such as social behaviors, controversial topics and customer oriented behaviors) could be detected. Moreover, customer satisfaction may be predicted.

Chapter 7, *A Linguistic Approach to Opinion Mining*, proposes an automatic linguistic approach to opinion mining. The proposed solution relies on a semantic analysis of textual resources and based on FreeWordNet, a new developed linguistic resource. FreeWordNet has been defined by the enrichment of the meanings expressed by adjectives and adverbs in WordNet with a set of properties and the polarity orientation. Reviews are used every day by common people or by companies who need to make decisions. Such amount of social data can be used to analyze the present and to predict the near future needs or the probable changes. Mining the opinions and the comments is a way to extract knowledge by previous experiences and by the feedback received.

Chapter 8, *Sentiment Analysis in the Planet Art: a Case Study in the Social Semantic Web*, set in a Social Semantic Web framework, explores the possibility of extracting rich, emotional semantic information from the tags freely associated to digitalized visual artworks, identifying the prevalent emotions that are captured by the tags. To this end, authors rely on ArsEmotica, an application software that combines an ontology of emotional concepts with available computational and sentiment lexicons. The Chapter reports and comments also results of a user study, aimed at validating the outcomes of ArsEmotica. Those results were obtained by involving the users of the same community which tagged the artworks.

In Chapter 9, *OntoTimeFL - A Formalism for Temporal Annotation and Reasoning for Natural Language Text*, an ontological formalism for annotating complex events expressed in natural language is defined. Compared to TimeFL, OntoTimeFL introduces new constructs in form of concepts for the annotation of three types of complex events: narrative, intentional, and causal events. In addition, the methodological choice of defining OntoTimeFL as a conceptualization of TimeFL makes easier the processes of automated annotation and of reuse and application of several types of axioms and rules for temporal reasoning to the annotated items.

Chapter 10, *Representing Non Classical Concepts in Formal Ontologies: Prototypes and Exemplars*, focuses on concept representation, an open problem in the field of ontology engineering and knowledge representation. Authors review empirical evidence from cognitive psychology, which suggests that concept representation is not an unitary phenomenon. In particular, they found that human beings employ both prototype and exemplar based representations in order to represent non classical concepts. Thus, authors suggest that a similar, hybrid prototype-exemplar based approach could be useful also in the field of formal ontology technology.

Chapter 11, *From Logical Forms to SPARQL Query with GETARUNS*, presents a Question Answering (QA) tool which integrates a full-fledged NLP system for text understanding, called GETARUNS. This system deals with different levels of syntactic and semantic ambiguity and generates some structures, called Logical Forms, by accessing computational lexical equipped with sub-categorization frames and appropriate selectional restrictions applied to the attachment of complements and adjuncts. Logical Forms are exploited by the QA system to compute a prospective answer and to extract the semantic elements needed to produce a SPARQL expression that is then used to query Linked Open Data endpoints.

Chapter 12, *A DHT-based Multi-Agent System for Semantic Information Sharing*, presents AOIS (Agents and Ontology based Information Sharing), a multi-agent system that supports the sharing of information among a dynamic community of users connected through BitTorrent, the well-known peer-to-peer platform. Compared to web scale search engines, AOIS enhances the search through domain ontologies, avoids the burden of publishing the information on the Web and guarantees a controlled and dynamic access to the information. Agent technologies are exploited to filter information coming from different users on the ground of the user previous experience, to propose new information that can be potentially interesting for a user in a push modality as well as to delegate access capabilities on basis of a reputation network built by the agents of the system on the user community.

Chapter 13, *A Decisional Multi-Agent Framework for Automatic Supply Chain Arrangement*, proposes a multi-agent system for supply chain dynamic configuration. Agent brain is composed of a Bayesian decision network, thus allowing agent to take the best decisions by estimating benefits and potential risks of different strategies, analyzing and managing uncertain information about the collaborating companies. In so doing, each agent collects information about customers orders and current market prices and is able to analyze previous experiences of collaborations with trading partners. Therefore, the agent performs a probabilistic inferential reasoning to filter information modeled in its knowledge base in order to achieve the best performance in the supply chain organization.

We would like to thanks all the authors for their excellent contributions and reviewers for their careful revision and suggestions for improving the proposals. We are grateful to the Springer-Verlag Team for their assistance during the preparation of the manuscript. We are also indebted to all participants and scientific committee members of the fifth edition of the DART workshop, for their continuous encouragement, support and suggestions.

May 2012   Cristian Lai
                                                                               Giovanni Semeraro
                                                                                  Eloisa Vargiu

Reviewers

Contents

A Brief Account on the Recent Advances in the Use of Quantum Mechanics for Information Retrieval

Massimo Melucci

Abstract. The key challenge of Information Retrieval (IR) is relevance, that is, the property of a document that makes the document relevant to the end user's information need in a given context. Relevance is a key challenge because it is an unknown property – were every document at every context marked with a relevance assessment prior to any information need or were it indexed so that a set of terms is assigned if and only if the document is relevant, retrieval would be perfect. Unfortunately, this is not the case. Although an IR system is in principle able to optimally rank documents, retrieval effectiveness depends on how the document collection has been represented, e.g. which terms are associated to which documents.

Quantum Theory (QT) can be used in IR because its mathematics describes the properties of documents and queries which would outperform the current retrieval models if these properties could be observed. Using QT in a different domain than Physics is not obvious, thus making the use of QT in IR is useful and in fact necessary.

To this end, we describe how to use QT in, and we explain how QT can improve the current situation of IR. In particular, we suggest how measurement uncertainty can be leveraged in IR for designing novel retrieval functions constrasting current retrieval functions which are based on the uncertainty generated at indexing time. Moreover, we describe IR in terms of signal detection. By viewing retrieval as signal detection, it is possible to cast some results of quantum signal detection to IR and define the procedures for designing an indexing algorithms which better represents documents than traditional algorithms. Lastly, we describe some experimental results that suggest the potential of QT in IR.

Massimo Melucci
University of Padua
e-mail: m.melucci@acm.org

C. Lai et al. (Eds.): New Challenges in Distributed Inf. Filtering and Retrieval, SCI 439, pp. 1–14.
springerlink.com   © Springer-Verlag Berlin Heidelberg 2013

1 Introduction

The IR problem can briefly be stated as follows: How can we retrieve as many relevant and as few non-relevant documents as possible given some properties about the documents stored in a collection? The problem is not trivial at all because the end users of the World Wide Web (WWW) experience loss in either recall[1] or precision[2] every day although research in IR dates back to the 1950s and already provided effective yet sub-optimal results that by now form sound state-of-the-art IR systems. The improvements obtained through IR models in the past cannot be more effective than those stated by the Probability Ranking Principle (PRP) when the parameters are accurately estimated [1]. The most effective techniques that can improve parameter estimation are in the realm of query expansion and relevance feedback. Some feedback mechanisms are surveyed in [2] and more recently in [3].

The key challenge of IR is relevance, that is, the property of a document that makes the document relevant to the end user's information need in a given context. There is a long history about relevance, context, problematic situations and all the related issues. The point is that relevance is central and no other discipline has it at its own basis. Good accounts are given in [4, 5, 6, 7]. Relevance is a key challenge because it is an unknown property – were every document at every context marked with a relevance assessment prior to any information need, retrieval would be perfect. Unfortunately, this is not the case. Although an end user can mark a visited document as relevant, users are reluctant to provide marks due to privacy or cognitive overload issues. Were the users willing to provide assessments the difficulty would not be overcome – relevance is context-dependent and retrieval is subject to uncertainty.

An IR system can decide about relevance only through observables. An observable is a property of an information object (e.g. a document, a query) that can be determined by some sequence of operations. Common observables are textual terms in IR yet other observables are becoming famous with the growing interest in other media: images; sound and music; video; "tweets"; tags. The operations that determine observables might involve indexing a document by means of an indexing algorithm and eventually storing index terms to an index. The point is that the observables must be effective, that is, enable the system to separate relevant documents from non-relevant documents as accurately as possible. Thus, if the system can observe effective observables, the unknownness of relevance can at least be partially overcome. Nevertheless, retrieval will be sub-optimal. A region of acceptance is the set of observed values for which relevance is accepted by the system with "to accept" meaning "to retrieve and display the document to the user". For example, a region of acceptance is the set of documents that contains a query

---

[1] This is the proportion of relevant documents that are retrieved.

[2] This is the proportion of retrieved documents that are relevant.

Table 1 The notation used in this deliverable for describing observables and states.

Notation	Observable and state
+	Relevance
−	Non-relevance
↑	Term presence
↓	Term absence
↑+	Term presence in relevant document
↑−	Term presence in non-relevant document
→+	Term absence in relevant document
→−	Term absence in non-relevant document

term. In general, a region of acceptance is combination of criteria and each criterion involves one or more observables.

Suppose that an observable is represented as vectors at a given angle and that an unknown property (e.g. relevance) is represented with either + or −. There are as many angles as the observed values. For example, if textual term occurrences are observed, one angle refers to presence, the other refers to absence. In the following we use the notation described in Table 1.

Suppose the region of acceptance is ↑ meaning that a document is accepted if and only if a term occurs in the document. The system's decision is

$$\uparrow \overset{?}{=} \begin{cases} \uparrow+ \text{ correct detection } P(\uparrow+) \text{ probability of detection} \\ \uparrow- \text{ false alarm } \quad P(\uparrow-) \text{ probability of false alarm} \end{cases}$$

where detection is the acceptance of relevant documents and false alarm is the acceptance of non-relevant documents.

Once a region of acceptance is established by means of a model, evaluation of this model is the next step. To this end prepare a test collection (e.g. seven documents) as a training set. Then, observe a term in the documents and assign either ↑ or → to mean presence or absence, respectively. An index would be

$$1\ 2\ 3\ 4\ 5\ 6\ 7$$
$$\uparrow\uparrow\rightarrow\uparrow\rightarrow\uparrow\uparrow$$

Afterward, collect relevance assessments (i.e. + or −):

$$1\quad 2\quad 3\quad 4\quad 5\quad 6\quad 7$$
$$\uparrow+\ \uparrow-\ \rightarrow+\ \uparrow+\ \rightarrow-\ \uparrow-\ \uparrow-$$

At retrieval time one may observe a query, for example either ↑ or →. The task is to decide for either acceptance or rejection. The effectiveness of the model can be measured as recall (fraction of detected relevant documents) and fallout (fraction of detected non-relevant documents). If the region of acceptance is ↑, recall is $\frac{2}{3}$ and fallout is $\frac{3}{4}$. In summary,

Region of acceptance	Recall	Fallout
Neither $\uparrow$ nor $\rightarrow$	0	0
$\uparrow$	$\frac{2}{3}$	$\frac{3}{4}$
$\rightarrow$	$\frac{1}{3}$	$\frac{1}{4}$
Both $\uparrow$ and $\rightarrow$	1	1

This table shows that (1) there are four possible regions of acceptance and (2) no region is uniformly best because recall and fallout are often positively correlated. Therefore, one has to resort to the Neyman-Pearson Lemma (NPL) which states that given the observed values from the documents of a collection, a system needs to pick out those values which make the ratio between recall and fallout not less than a threshold, with fallout being not greater than a given level [8]. In practice, if fallout must not be greater than $\frac{1}{4}$, the best region of acceptance is $\rightarrow$; if fallout must not be greater than $\frac{3}{4}$, the best region of acceptance is $\uparrow$; if fallout must not be greater than 1 (i.e. any fallout), the best region of acceptance is both $\uparrow$ and $\rightarrow$ (i.e. all the documents).

Suppose that each region corresponds to a position (e.g. $\uparrow$ or $\rightarrow$). In this view, an IR model behaves like a detector with a fixed number of positions corresponding to the possible regions of acceptance. Yet with probability very near to zero, perfect retrieval (fallout $\frac{0}{4}$, recall $\frac{3}{3}$) can be achieved:

$$\uparrow+ \ \rightarrow- \ \uparrow+ \ \uparrow+ \ \rightarrow- \ \rightarrow- \ \rightarrow- \ \boxed{\uparrow} \ \uparrow+ \ \uparrow+ \ \uparrow+$$

where the box refers to a detector that takes information about term occurrence as input and produces a decision as output. The symbol inside the box refers to the angle at which the detector has been put. If the angle is zero (i.e. horizontal arrow) every horizontal arrow is accepted, every vertical arrow is rejected, and every oblique arrow is randomly accepted with probability provided by the squared amplitude of the complex number resulting from the product between the oblique arrow and the horizontal arrow. In most cases, retrieval is imperfect (e.g. given fallout $\frac{3}{4}$, recall $\frac{2}{3}$), e.g.

$$\uparrow+ \ \uparrow- \ \rightarrow+ \ \uparrow+ \ \rightarrow- \ \uparrow- \ \uparrow- \ \boxed{\uparrow} \ \uparrow+ \ \uparrow- \ \uparrow+ \ \uparrow- \ \uparrow-$$

However, retrieval may be optimal thanks to the NPL (e.g. given fallout $\frac{1}{2}$, recall is $\frac{1}{3}$):

$$\uparrow+ \ \uparrow- \ \rightarrow+ \ \uparrow+ \ \rightarrow- \ \uparrow- \ \uparrow- \ \boxed{\rightarrow} \ \rightarrow+ \ \rightarrow-$$

These examples clearly show what causes uncertainty (and error): the lack of one:one correspondence between an observable (e.g. relevance) and another observable (e.g. term occurrence).

2 Quantum Probabilility and IR

According to QT, not only a system can be set at one state out of a predefined set of states but it can also be set at superposed states – in classical
probability an elementary event (e.g. a ball in an urn) must be at one and
only one state (e.g. the ball must be either white or black) and there cannot
be superposed states (e.g. the ball is both white and black yet it is neither
gray). Borrowing these concepts in classical IR, documents must be at one
and only one state (i.e. term occurrence) chosen from a set of predefined values (e.g. term presence, term absence) and there cannot be superposed states
(e.g. term present and term absent in the same moment of time).

Superposed term occurrence is radically different from weighted term occurrence. Term occurrence is a binary representation and has been used
mainly in the early IR systems. Modern IR system represents term with a
weight of continuation value (e.g. TF.IDF representation), which can achieve
better performance in comparison with the binary representation. However,
such a weight is just an attribute of a binary event and does not at all represent another event. The use of a weight such as TF.IDF implies that the term
has occurred. Therefore, the weighting schemes adopted by IR systems are
not surrogates of the theory illustrated in this document. When either term
occurrence or non-occurrence representation is considered as it is customary
in IR superposition cannot be represented and a QT style representation is
necessary. This approach is addressed in the remainder of this section.

Suppose the detector can be set to different positions than those fixed,
that is, it can be "angled" in between. Angling is the key point, it is not
(yet) possible in classical IR and is neither (yet) possible in every field of
Computer Science that requires observing the values building regions of acceptance made from sets of these values. For example, if it is *a priori* decided
that a term either occurs or does not, it is not possible *a posteriori*, once it
has already been observed, to say that maybe it occurs and to represent this
event using an angled vector, that is a vector angled in between.

When the vector representing the detector is angled, measurement adds
uncertainty. Each input document that is either → or ↑ reaches the detector
set at an angled configuration and is randomly filtered by the detector. Indeed, when a document reaches a detector like $\boxed{\nearrow}$, at detection time, the
document passes through the detector and changes to either → or ↑ according to a probability distribution given by the Born rule. Suppose a detector
has to be angled so that we have for example:

$$↑+ \; ↑- \; →+ \; ↑+ \; →- \; ↑- \; ↑- \; \boxed{\nearrow} \; →+ \; ↑- \; ↑+$$

In this example, the symbols on the right hand side represent relevant or
non-relevant documents. These documents are randomly filtered as ↑ or a
→ when passing through $\boxed{\nearrow}$. One can check that recall is $\frac{1}{2}$ because one
relevant document has been retrieved out of two relevant input documents.

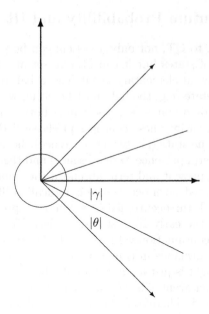

Fig. 1 An example of construction of the optimal detector. Suppose $P(\rightarrow+) = \frac{1}{3}, P(\rightarrow-) = \frac{1}{2}$. Then $|X|^2 = \sqrt{\frac{2}{3}}$. And $\theta = \frac{1}{2}\left(\frac{\pi}{2} - \arccos|X|^2\right) \approx 0.37$. Compute $\gamma = \arccos\sqrt{P(\rightarrow+)} \approx 1.23$. Rotate $\rightarrow$ by θ, γ.

The question is: How can we angle the detector so that recall is greater than $\frac{1}{2}$ given the same fallout?

There is an optimality result that ensures that an optimal angle can be found. Suppose the recall values of the regions of acceptance represented by $\rightarrow$ and $\nearrow$ are $P(\rightarrow+)$ and $P(\nearrow+)$, respectively. Similarly, the fallout values are, respectively, $P(\rightarrow-)$ and $P(\nearrow-)$. It can be proved that there exists an optimal angle such that $P(\nearrow+)$ is always greater than or equal to $P(\rightarrow+)$ for every $P(\rightarrow-) = P(\nearrow-)$. This result is proved in [9] for IR, yet it derives from the theory of Quantum Detection (QD) described in [10].

The optimal detector $\boxed{\nearrow}$ can be built as follows:

- Take $P(\uparrow+), P(\uparrow-), P(\rightarrow+) = 1 - P(\uparrow+), P(\rightarrow-) = 1 - P(\uparrow-)$.
- Compute $|X|^2 = \sqrt{P(\uparrow+)P(\uparrow-)} + \sqrt{P(\rightarrow+)P(\rightarrow-)}$; $|X|^2$ is the best distance function between two probability distributions [11].
- Compute $\theta = \frac{1}{2}\left(\frac{\pi}{2} - \arccos|X|^2\right)$.
- Compute $\gamma = \arccos\sqrt{P(\rightarrow+)}$.
- Rotate $\rightarrow, \uparrow$ by θ and then by γ.

An example is in Figure 1.

The optimal detector is a mathematical representation of what should actually be observed instead of what is observed, for improving recall at every fallout. Indeed, the procedure for building the optimal detector first assumes that an observable $\rightarrow, \uparrow$ is selected, then, the probabilities are estimated and lastly the optimal detector is found. No hints are given for what to observe from the documents. This situation can be exemplified as:

$$\uparrow+ \;\uparrow- \;\rightarrow+ \;\uparrow+ \;\rightarrow- \;\uparrow- \;\uparrow- \;\boxed{\nearrow} \;\uparrow+ \;\uparrow- \;\rightarrow+ \;\uparrow+ \;\uparrow-$$

In this example, the resulting documents match the optimal detector and are described by either $\rightarrow$ or $\uparrow$. Are situations like the example always possible? No, they are not.

Two observables $\nearrow, \searrow$ and $\rightarrow, \uparrow$ might not be observed together, i.e. the simultaneous occurrence of two descriptors in a document collection must be subject to some constraints. If these constraints are not admitted in a collection, this simultaneous occurrence cannot be admitted either.

In [12] it has been shown that observables like $\nearrow, \searrow$ and $\rightarrow, \uparrow$ can be estimated by a single sample space if and only if some Statistical Invariants (SIs) hold. When there are two observables, the SI is given by:

$$\left| \frac{P(\nearrow) - P(\nearrow|\uparrow)}{P(\nearrow|\rightarrow) - P(\nearrow|\uparrow)} \right| \leq 1 \tag{1}$$

Other SIs are given in [13, 14], for example.

This SI means that $\nearrow, \searrow$ and $\rightarrow, \uparrow$ can be observed together if and only if it holds, given that the probabilities $P(\nearrow), P(\nearrow|\uparrow), P(\nearrow|\rightarrow)$ are as accurately estimated as possible although these probabilities might not be estimated from only one source of evidence.

3 Experimentation

We conducted three different experiments in the recent years. These experiments differ in terms of test collection, aims and methodology. They are reported in [15], [16], [9] and, respectively, are summarized as follows:

1. Simulation: Were a query term the optimal detector, what would its performance be.
2. Estimation: Rank query expansion terms by the degree to which they violate the SIs.
3. Observation: Observe human subjects interacting with documents and predict relevance using SIs.

In the simulation, the disks 4 and 5 of the TIPSTER test collection (about 500,000 fulltext documents) and the TREC 6, 7, 8 topic sets were used – topic title-only queries were submitted to the experimental system. The following probabilities of error were used to measure and compare retrieval effectiveness:

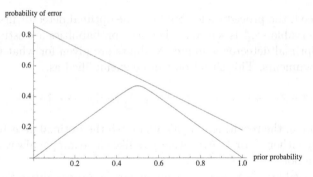

Fig. 2 P_e and Q_e plotted against $P(-)$ for term `crime` of topic 301. Q_e corresponds to the bell-shaped curve. Q_e is always greater than or equal to P_e for every $P(-)$.

$$P_e = P(\rightarrow +) + P(\uparrow -)$$
$$Q_e = P(\searrow +) + P(\nearrow -)$$

where $\uparrow +$ is supposed to be the best region of acceptance and $+$ means relevance. For each query term, the probability of occurrence in relevant documents and the probability of occurrence in non-relevant documents have been estimated according to IR literature [17]:

$$P(\uparrow +) = \frac{r}{R} \qquad P(\uparrow -) = \frac{n - r}{N - R}$$

where r is the number of relevant documents indexed by the query term out of the total R relevant documents and n is the number of documents indexed by the query term out of the total N documents. For example, the query term `crime` of topic 301 has the following probabilities:

$$P(\uparrow +) = \frac{65}{474} \qquad P(\uparrow -) = \frac{223}{1234}$$

thus yielding $|X|^2 = 0.998217$ – the probability distributions are very close to each other. The probabilities of error are plotted against $P(-)$ for term `crime` of topic 301 in Figure 2.

In the second experiment, the same test collection was used. In addition, the 32 best runs from each of the set of runs submitted to TREC-6, TREC-7 and TREC-8 were selected – these best runs represent the state-of-the-art at that time. For each run, every term $\uparrow$ from the top n documents has been extracted and classified according to either $+$ or $-$. Given $n(x)$ the number of documents with x within the top n and $N(x)$ the number of documents with x within the whole collection,

$$P(\uparrow +) = \frac{n(\uparrow +)}{n(+)} \qquad P(\uparrow -) = \frac{n(\uparrow -)}{n(-)} \qquad P(\uparrow) = \frac{N(\uparrow)}{N}$$

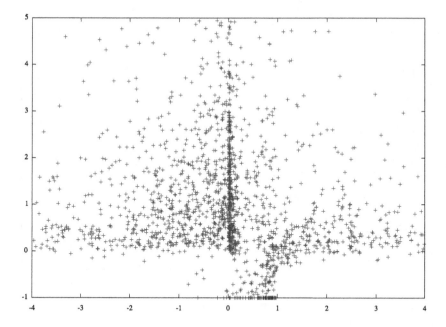

Fig. 3 The scatterplot of SI_n, Δ_n for $n = 20$ and manual query expansion.

Then, SI_n was computed for each term and within the top n documents, i.e.

$$SI_n(\uparrow) = \left| \frac{P(\uparrow) - P(\uparrow-)}{P(\uparrow+) - P(\uparrow-)} \right|$$

Moreover, for every term, the whole collection was re-ranked by the term and the variation in recall within the top n documents has been computed. In this way, SI_n has been mapped to

$$\Delta_n(\uparrow) = \frac{P(\uparrow) - P(\uparrow+)}{P(\uparrow+)}$$

and the scatterplots of Figures 3 and 4 were achieved. In both scatterplots, a dense concentration of points in the $[0, 1]$ interval of SI can be observed and, in particular, associated with both negative and positive values of Δ. Outside this interval, Δ is more frequently positive than negative, thus meaning that the terms that violate the SI tend to be more effective than the terms that do not. These results suggest that the SI is a means of detecting terms which are effective in increasing recall and the overall retrieval performance.

The third experimentation aimed at measuring the interference term. Through a user study, the subjects were asked to access a document collection and to assess the relevance of the displayed documents. The accesses and the assessments were logged. In particular, display time (i.e. the time

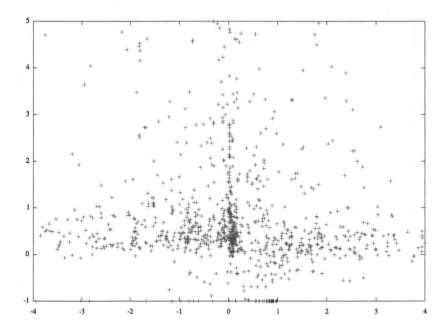

Fig. 4 The scatterplot of SI_n, Δ_n for $n = 20$ and automatic query expansion ($k = 10$).

a page was on focus) was logged. In this way, it was possible to count the number of accesses of a user to a page by dividing the display time into constant intervals. For example, the user `user1` interacted with a page and made the relevance assessment out of four possible grades after 167 seconds, he interacted with the page in 17 distinct and consecutive intervals and the assessment was recorded only at the 17th interval.

If another user was interacting with the same document (or the same user was interacting with another document) for 16 seconds, another distinct event is recorded and the display time of 16 seconds is divided by the predefined width by giving rise to, say, two new intervals – the first interval is ten seconds long and the second interval is six seconds long which after another assessment was made and the interaction stopped. In total, after two events, there would be two accesses in the first ten-second interval, two accesses in the second ten-second interval, and one access in each other interval.

The results are reported in [18]. It was found that the interference term is significant at the early intervals when the proportion of accesses was the highest and, incidentally, when the user's interaction often ends with an assessment. This outcome signals that even when it is supposed that the majority of users reach an agreement on relevance, interference in the user's relevance state may occur. Some other remarks about these results are reported in [18].

4   Related Work

QT is a theory of Physics, yet it has recently been studied in other fields such as: Computer Science; Biology; Cognition and Psychology; Information Access and Retrieval; Natural Language Processing; Economics and Finance; Social Studies. The Quantum Interaction symposia have been gathering researchers from these areas since 2007. In particular, IR and QT have been studied since the book published in [19]. A recent survey is in [20]. Moreover, some popular articles have been published[3] and some research projects have started[4,5].

The foundations of QT have been illustrated in numerous books such as [21] and [22]: the former is especially interesting for the researchers with a background in Statistics who would like to enter into QT; the latter explains the subject from a variety of perspectives (i.e. logical, geometrical, and probabilistic). Quantum probability, for example, has been introduced in [23]. In particular, the interference term is addressed in [12] where the concept of Statistical Invariant (SI) is introduced. The utilization of QT in computation, information processing and communication is described in [24], which by now is a classical text. Recently, investigations have started in other research areas, for example, in IR [25]. An up-to-date account of QT in IR is in [20].

The optimal detector illustrated in this deliverable is inspired by Helstrom's book [10] which provides the foundations and the main results in QD; an example of the exploitation within communication theory of these results of QD is reported [26]. This deliverable is founded on the application of QD to IR firstly reported in [9], yet it links to [19] with regard to quantum probability.

In IR, the Probability Ranking Principle (PRP) states that "If a reference retrieval system's response to each request is a ranking of the documents in the collection in order of decreasing probability of relevance to the user who submitted the request, where the probabilities are estimated as accurately as possible on the basis of whatever data have been made available to the system for this purpose, the overall effectiveness of the system to its user will be the best that is obtainable on the basis of those data" [1]. However, the assumption that the optimal detector can be set at fixed positions undermines the general applicability of the PRP. The results of this deliverable depart from the PRP for IR since we enhance the principle by allowing the optimal detector to be fixed at other positions.

In [27], the authors propose ranking documents by quantum probability and suggests that interference (which must be estimated) might model dependencies between relevance judgements such that documents ranked until

---

[3] http://www.newscientist.com/article/mg21128285.900-quantum-minds-why-we-think-like-quarks.html

[4] http://qontext.dei.unipd.it

[5] http://renaissance.dcs.gla.ac.uk

position $n-1$ interfere with the degree of the relevance of the document ranked at position n.

In [28], the authors discuss how to employ quantum formalisms for encompassing various IR tasks within only one framework. From an experimental point of view, what that paper demonstrates is that ranking functions based on quantum formalism are computationally feasible. As the quantum formalism is mainly used to express IR tasks, the best experimental results of rankings driven by quantum formalism are comparable to BM25, i.e. to the PRP, thus limiting the contribution within a classical probability framework.

The study of the presence of quantum phenomena in IR and in general the evaluation of quantum-like models are still at the beginning. Nevertheless, there are already some efforts which may be classified into two main classes. The first class of experiments includes the classical experiments aiming at comparing the quantum-like models with the non-quantum-like ones which are used as baseline for measuring the difference in precision; examples of the first class can be found in [29, 30]. The second class of experiments includes the experiments designed for testing the hypothesis that quantum phenomena in processes of IR or of areas related to IR occurs. As regards the areas related to IR, the experiments have mainly addressed the presence of interference in human cognition and lexicon and have been based on user studies; examples are reported in [31, 32, 33]. An experiment more drectly related to IR is reported in [34]; it discusses a situation in which quantum probability arises naturally in IR, and reports that the best terms for query expansion have probabilities which do not admit classical probability but instead can be defined by a quantum probability function.

5   Conclusions

The improvements obtained through IR models in the past cannot be more effective than those stated by the PRP under the assumptions stated in [1] and the fact that a region of acceptance is based on predefined observables. We have conjectured that computing alternative observables is in principle more effective than designing new techniques or term weighting schemes.

We have proved that a significant improvement can in principle be attained if the detector that represent a retrieval function is "angled", but we need to build such a detector. The design of such a detector is not trivial at all because the "angled" symbols do not straightforwardly correspond to the "physical" events (e.g. feature occurrence) with which IR systems deal.

Acknowledgements. This work has received funding from the European Union Seventh Framework Programme (FP7/2007-2013) under grant agreement N. 247590 for the Marie Curie IRSES project "Quantum Contextual Information Access and Retrieval" (QONTEXT).

References

1. Robertson, S.: The probability ranking principle in information retrieval. Journal of Documentation 33, 294–304 (1977)
2. Lalmas, M., Ruthven, I.: A survey on the use of relevance feedback for information access systems. Knowledge Engineering Review 18 (2003)
3. Carpineto, C., Romano, G.: A survey of automatic query expansion in information retrieval. ACM Comput. Surv. 44, 1:1–1:50 (2012)
4. Ingwersen, P., Järvelin, K.: The Turn: Integration of Information Seeking and Retrieval in Context. Springer (2005)
5. Mizzaro, S.: Relevance: The whole history. Journal of the American Society for Information Science 48, 810–832 (1997)
6. Saracevic, T.: Relevance: A review of the literature and a framework for thinking on the notion in information science. Part II: nature and manifestations of relevance. Journal of the American Society for Information Science and Technology 58, 1915–1933 (2007)
7. Saracevic, T.: Relevance: A review of the literature and a framework for thinking on the notion in information science. Part III: Behavior and effects of relevance. Journal of the American Society for Information Science and Technology 58, 2126–2144 (2007)
8. Neyman, J., Pearson, E.: On the problem of the most efficient tests of statistical hypotheses. Philosophical Transactions of the Royal Society, Series A 231, 289–337 (1933)
9. Melucci, M.: Can Information Retrieval Systems Be Improved Using Quantum Probability? In: Amati, G., Crestani, F. (eds.) ICTIR 2011. LNCS, vol. 6931, pp. 139–150. Springer, Heidelberg (2011)
10. Helstrom, C.: Quantum Detection and Estimation Theory. Academic Press, New York (1976)
11. Wootters, W.K.: Statistical distance and Hilbert space. Phys. Rev. D 23, 357–362 (1981)
12. Accardi, L.: On the probabilistic roots of the quantum mechanical paradoxes. In: Diner, S., de Broglie, L. (eds.) The Wave-Particle Dualism, pp. 297–330. D. Reidel pub. co. (1984)
13. Accardi, L., Fedullo, A.: On the statistical meaning of complex numbers in quantum mechanics. Lettere al Nuovo Cimento 34, 161–172 (1982)
14. Pitowsky, I.: Quantum Probability – Quantum Logic. Springer (1989)
15. Melucci, M.: An Investigation of Quantum Interference in Information Retrieval. In: Cunningham, H., Hanbury, A., Rüger, S. (eds.) IRFC 2010. LNCS, vol. 6107, pp. 136–151. Springer, Heidelberg (2010)
16. Di Buccio, E., Melucci, M., Song, D.: Towards Predicting Relevance Using a Quantum-Like Framework. In: Clough, P., Foley, C., Gurrin, C., Jones, G.J.F., Kraaij, W., Lee, H., Mudoch, V. (eds.) ECIR 2011. LNCS, vol. 6611, pp. 755–758. Springer, Heidelberg (2011)
17. van Rijsbergen, C.: Information Retrieval, 2nd edn. Butterworths, London (1979), Disponibile in formato elettronico ipertestuale all'indirizzo http://www.dei.unipd.it/~melo/htb/
18. Di Buccio, E., Melucci, M., Song, D.: Exploring combinations of sources for interaction features for document re-ranking (2010), http://research.microsoft.com/en-us/um/people/ryenw/hcir2010/docs/HCIR2010Proceedings.pdf

19. van Rijsbergen, K.: The geometry of information retrieval. Cambridge University Press, UK (2004)
20. Melucci, M., van Rijsbergen, K.: Quantum mechanics and information retrieval. In: Advanced Topics in Information Retrieval, pp. 125–155. Springer, Berlin (2011)
21. Griffiths, R.B.: Consistent quantum theory. Cambridge University Press, Cambridge (2002)
22. Hughes, R.: The structure and interpretation of quantum mechanics. Harvard University Press, Cambridge (1989)
23. Parthasarathy, K.: An Introduction to Quantum Stochastic Calculus, Birkhäuser (1992)
24. Nielsen, M., Chuang, I.: Quantum Computation and Quantum Information. Cambridge University Press (2000)
25. Bruza, P., Kitto, K., Nelson, D., McEvoy, C.: Extracting Spooky-Activation-at-a-Distance from Considerations of Entanglement. In: Bruza, P., Sofge, D., Lawless, W., van Rijsbergen, K., Klusch, M. (eds.) QI 2009. LNCS, vol. 5494, pp. 71–83. Springer, Heidelberg (2009)
26. Cariolaro, G., Pierobon, G.: Performance of quantum data transmission systems in the presence of thermal noise. IEEE Transactions on Communications 58, 623–630 (2010)
27. Zuccon, G., Azzopardi, L.A., van Rijsbergen, K.: The Quantum Probability Ranking Principle for Information Retrieval. In: Azzopardi, L., Kazai, G., Robertson, S., Rüger, S., Shokouhi, M., Song, D., Yilmaz, E. (eds.) ICTIR 2009. LNCS, vol. 5766, pp. 232–240. Springer, Heidelberg (2009)
28. Piwowarski, B., Frommholz, I., Lalmas, M., van Rijsbergen, K.: What can quantum theory bring to information retrieval? In: Proc. 19th International Conference on Information and Knowledge Management, pp. 59–68 (2010)
29. Huertas-Rosero, A., Azzopardi, L., van Rijsbergen, K.: Eraser lattices and semantic contents: An exploration of the semantic contents in order relations between erasers. In: [35], pp. 266–275
30. Zuccon, G., Azzopardi, L.: Using the quantum probability ranking principle to rank interdependent documents. In: Proceedings of the European Conference on Information Retrieval, pp. 357–369 (2010)
31. Aerts, D., Gabora, L.: A theory of concepts and their combinations II: A Hilbert space representation. Kybernetes 34, 176–205 (2005)
32. Busemeyer, J.R.: Introduction to Quantum Probability for Social and Behavioral Scientists. In: Bruza, P., Sofge, D., Lawless, W., van Rijsbergen, K., Klusch, M. (eds.) QI 2009. LNCS, vol. 5494, pp. 1–2. Springer, Heidelberg (2009)
33. Hou, Y., Song, D.: Characterizing pure high-order entanglements in lexical semantic spaces via information geometry. In: [35], pp. 237–250
34. Melucci, M.: An Investigation of Quantum Interference in Information Retrieval. In: Cunningham, H., Hanbury, A., Rüger, S. (eds.) IRFC 2010. LNCS, vol. 6107, pp. 136–151. Springer, Heidelberg (2010)
35. Bruza, P., Sofge, D., Lawless, W., van Rijsbergen, K., Klusch, M. (eds.): QI 2009. LNCS, vol. 5494. Springer, Heidelberg (2009)

Cold Start Problem: A Lightweight Approach

Leo Iaquinta, Giovanni Semeraro, and Pasquale Lops

Abstract. The chapter presents the SWAPTeam[1] participation at the ECML/PKDD 2011 - Discovery Challenge for the task on the cold start problem focused on making recommendations for new video lectures. The developed solution uses a content-based approach because it is less sensitive to the cold start problem that is commonly associated with pure collaborative filtering recommenders. The Challenge organizers encouraged solutions that can actually affect VideoLecture.net, thus the proposed integration strategy is the hybridization by switching. In addition, the surrounding idea for the proposed solution is that providing recommendations about cold items remains a chancy task, thus a computational resource curtailment for such task is a reasonable strategy to control performance trade-off of a day-to-day running system. The main contribution concerns about the compromise between recommendation accuracy and scalability performance of proposed approach.

1 Introduction

The chapter presents the approach proposed for the ECML/PKDD 2011 - Discovery Challenge[2] [6] and revised after the Discovery Challenge Workshop, where 9 teams (out of the 62 teams active on task 1 and the 22 teams active on task 2) have presented and discussed alternative solutions.

Leo Iaquinta
University of Milano–Bicocca, v.le Sarca 336, 20126 Milano, Italy
e-mail: iaquinta@disco.unimib.it

Giovanni Semeraro · Pasquale Lops
University of Bari "Aldo Moro", v. Orabona 4, 70125 Bari, Italy
e-mail: {semeraro, lops}@di.uniba.it

[1] http://www.di.uniba.it/~swap/index.php
[2] http://www.ecmlpkdd2011.org/challenge.php

C. Lai et al. (Eds.): New Challenges in Distributed Inf. Filtering and Retrieval, SCI 439, pp. 15–32.
springerlink.com   © Springer-Verlag Berlin Heidelberg 2013

The first of the two tasks of the ECML/PKDD 2011 - Discovery Challenge was focused on the cold start problem. Cold start is commonly associated with pure collaborative filtering-based Recommender Systems (RSs). RSs usually suggest items of interest to users by exploiting explicit and implicit feedbacks and preferences, usage patterns, and user or item attributes. Past behaviour is assumed to be useful to make reliable predictions, thus past data is used in the training of RSs to achieve accurate prediction models. Particularly, item-based collaborative filtering techniques assume that items are similar when they are similarly rated and therefore the recommendations concern about items with the highest correlations according to the usage evidence. A straight drawback is that new items cannot be recommended because there is not an adequate usage evidence. As a consequence, recommendations may be negatively affected by the well-known *cold start* problem. A design challenge comes from the dynamism of real-world systems because new items and new users whose behaviour is unknown are continuously added into the system.

The ECML/PKDD 2011 - Discovery Challenge dataset was gathered from VideoLectures.Net web site. VideoLectures.Net exploits a RS to guide users during the access to its large multimedia repository of video lectures. Beside the editorial effort to select and classify lectures, accompanying documents, information and links, the Discovery Challenge is organized in order to improve the website's current RS, inter alia, to deal with the cold start problem.

The main idea underlying our participation is to use a content-based approach because it is less sensitive to the cold start problem. Indeed, to overcome the cold start problem in the approaches based on collaborative filtering, a common solution is to hybridize them with techniques that do not suffer from the same problem [2]. Thus, a content-based approach is used to bridge the gap between existing items and new ones: item attributes are used to infer similarities between items. The adopted solution exploits almost all the provided data and the actual integration with VideoLectures.Net RS can be potentially performed by a hybrid approach.

Moreover, the scalability performance is considered as a primary requirement and, thus, a lightweight solution is pursued. In addition, evaluating the system accuracy on cold items it may be wise to consider that there is a trade-off with the entire system accuracy [9]. Because predictions involving *cold* items are more troublesome than *hot* items, obtaining recommendations about cold items by modest computational resources allows to encourage novelty and serendipity without penalize the entire system performances.

The rest of the chapter is structured as follows: Section 2 recalls some common knowledge about the cold start problem, Section 3 sketches some features of the dataset, Section 4 introduces the evaluation metric, Section 5 illustrates the proposed solution and Section 6 closes the chapter with some conclusions and future work.

2 Cold Start Problem

The cold start problem is commonly associated with pure collaborative filtering RSs. Collaborative filtering techniques [10] analyse interactions between all users and all items through user behaviour, i.e. rating, clicking, commenting, tagging or buying. Collaborative filtering RSs do not use any specific knowledge about items except their unique identifiers and thus they are domain-independent. However, collaborative filtering needs sufficient amount of collaborative data to provide recommendation to a new user or about a new item (the cold-start problem) [8].

The cold start problem concerns performance issues when new items (or new users) should be handled by the system. The cold start can be considered as a sub problem of the coverage one [9], indeed it measures the system coverage over a specific set of items or users. Therefore, although the prediction accuracy of a RS, especially for a collaborative filtering one, often grows with the amount of data, the coverage problem of some algorithms appears with recommendations of high quality only for a portion of the items even if the system has gathered a huge amount of data.

Focusing on cold start for items, there are various heuristics to pick out the cold items. For instance, cold items can be items with no ratings or usage evidence, or items that exist in the systems for less than a certain amount of time (e.g., a day), or items that have less than a predefined evidence amount (e.g., less than 10 ratings) [8]. The selection strategy is a parameter of the overall recommendation problem. Section 3 briefly introduces the selection strategy devised by the Challenge organizers for simulation scenario in the task 1. The correct selection of cold items allows to process them in a different way. Indeed the lack of homogeneity on amount of available data for cold and hot items causes issues on overall performance when a single approach is used to obtain recommendations about both cold and hot items, for instance recommendations can be more accurate about hot items than ones about cold items. In addition, boosting cold items without any meaningful criterion should promote the gathering of data about them, but such strategy should prejudice the user trust on all recommendations.

Moreover a common expectation for recommender systems concerns with novelty and serendipity, even if they cannot suitably measured by traditional accuracy metrics. Thus evaluating the system accuracy on cold items it may be wise to consider that there is a trade-off with the entire system accuracy [9] and cold items can promote novelty and serendipity in recommendations.

3 Dataset

The Challenge dataset is a data snapshot from the VideoLectures.Net taken in August 2010. At that time, the database contained 8,105 video lectures.

The main entities of the dataset are the lectures. They are described by a set of attributes and relationships. The attributes are of various kind: for instance, *type* can have only one value in a predefined set (lecture, keynote, tutorial, invited talk and so on); *views* attribute has a numeric value; *rec_date* and *pub_date* have a date value; *name* and *description* are unstructured text, usually in the language of the lecture. The relationships link the lectures with 519 context events, 8,092 authors, and 348 categories. Each of these entities has its own attributes and relationships to describe taxonomies of events and categories.

Almost all this amount of data can be exploited to obtain features for a content-based recommendation approach. The used features are briefly introduced in Section 5.2.

In addition, the dataset contains records about pairs of lectures viewed together (not necessarily consecutively) with at least two distinct cookie-identified browsers. This kind of data has a collaborative flavour and it is actually the only information about the past behavior. The user identification is missing, thus any user personalization is eliminated. User queries and feedbacks are also missing.

The whole dataset is split into 6,983 lectures for the training and 1,122 lectures for the testing as cold items. The splitting is performed considering the undirected weighted graph G of co-viewed lectures. Each lecture in G has associated a temporal information about the date of publishing at the VideoLectures.Net site and G is partitioned into two disjoint graphs G_1 and G_2 using a threshold on publishing date by the threshold. Each lecture in G_1 has the publishing date before the date threshold while each lecture in G_2 has the publishing date after the date threshold.

To obtain similar distribution for the *pair common viewing times* (PCVT), i.e. the period that two lectures spend together in the system, in both training and test sets, the Challenge organizers have randomly divided nodes from subgraph G_2 into two approximately equal sets (G_{21}, G_{22}) and then they have appended G_{21} to the training set. Thus, the subset of lecture pairs $(x_i, x_j) : x_i \in G_1, x_j \in G_{21}$ from the training set has similar distribution of PCVTs that overlaps with times of $(x_i, x_j) : x_i \in G_1, x_j \in G_{22}$ from the test set. More details are reported in [1].

4 The Cold Start Task and Evaluation Metric

Besides the quality of recommendations should be profitably measured through user satisfaction surveys and analysis, a challenge execution needs a quantitative measure in order to score solutions to a simulated situation. Thus, the first task of ECML/PKDD 2011 - Discovery Challenge assumed that a user receives suggestions about *new* lectures after has seen one of the *old* lectures, i.e. one of the lectures which are characterized by the earlier times of entering in the system in contrast to those lately added. The length of the recommended list is fixed at 30 lectures.

Overall score is based on the mean average R-precision score (*MARp*) as a variant of standard evaluation measures in information retrieval $p@k$ and *MAP*. More formally, the overall score of a solution is the mean value over all queries R in the test set:

$$MARp = \frac{1}{|R|} \sum_{r \in R} AvgRp(r)$$

where $AvgRp(r)$ is the average R-precision score for the recommended ranked list r obtained for an old item and it is defined as:

$$AvgRp(r) = \sum_{z \in Z} \frac{Rp@z(r)}{|Z|}$$

where $Rp@z(r)$ is R-precision at some cut-off length $z \in Z$ and it is defined as:

$$Rp@z(r) = \frac{|relevant \cap retrieved|_z}{|relevant|_z} = \frac{|relevant \cap retrieved|_z}{\min(m,z)}$$

Number of relevant items at cut-off length z is defined as $\min(m,z)$, where m is the total number of relevant items. For the task 1, cut-off lengths z for the calculation of $MARp$ are $z \in \{5, 10, 15, 20, 25, 30\}$.

The organizers have devised two version of $MARp$. The preliminary version is mainly used during the Challenge to provide feedback to the limited submissions of the participants. The final versions is used to draw up the Challenge graded list. In the task 1, the majority of teams have positive differences between preliminary $MARp$ score on the leaderboard set and the final $MARp$ score on the test set, which may suggest overtraining. More details are reported in [1].

The scores of the provided baseline random solution are 0.01949 (preliminary) and 0.01805 (final).

5 Proposed Approach

5.1 Content-Based Technique by Hybrid Approach

To overcome the cold start problem of the collaborative approaches, a common solution is to hybridize them with other techniques that do not suffer of the same problem [2]. For instance, a content-based approach can be used to bridge the gap from existing items to new ones: item attributes are used to infer similarities between items.

Content-based techniques also have a start-up problem because they must accumulate enough usage evidence to build a reliable classifier, but in the task on the cold start problem of the ECML/PKDD 2011 - Discovery Challenge it is not an issue because the organizers focused the task on *item-to-item* recommendations for already started system and, consequently, many issues about start-up for content-based approaches are cut-off.

Furthermore, relative to collaborative filtering, content-based techniques are limited by the features that are explicitly associated with the items that they recommend. For instance, a content-based movie recommendation is usually based on the movie metadata, since the movie itself is opaque to the system. In the task on the cold start problem of the ECML/PKDD 2011 - Discovery Challenge, this general problem is solved by the editorial effort of VideoLectures.Net to select and classify lectures. In addition, as sketched in Section 3, almost all provided data can be exploited to obtain content-based features.

The hybridization strategy can be flexible in order to apply different approaches to specific classes of items (or users) and, therefore, switch to a specific technique for the selected cold items. A switching approach [2] is a simple hybridization strategy to implement different techniques with sensitivity on the item-level without any further cost beside the cold item selection.

5.2 Steps towards Solution

The solution is obtained mainly by three steps: the data pre-processing, the model learning, and the recommendation.

Data pre-processing step starts by obtaining an in-memory object-oriented representation. In addition, a set of language-dependant indexes are created to store textual metadata (title, description and slide title) in order to exploit the term frequency vectors to efficiently compute document similarities. The textual metadata is also preprocessed to remove stop words and to reduce inflected words to their stem: these sub-steps are strongly language-dependent, thus specific linguistic knowledges can improve the process effectiveness. The event names are filtered by regular expressions to introduce an event similarity metric smarter than a simple string matching. An in-memory complete representation of category taxonomy is also created to compute the category similarity as graph-based minimum path between pairs of categories.

The main output of this step is a set of 19 numeric values describing the similarities between lectures of each pair in the training set. Table 1 reports the used features: for each pair of items, they involve the languages, the frequencies of languages (Fig. 8-b), the descriptions, the recording and publication ages, the conferences, the authors and their affiliations, and the categories.

Model learning step uses the popular machine learning suite Weka[3] to build a prediction model for the frequency of a pair of lectures. The available data and the lightweight goal determined the selection of a linear model for the learning problem. Thus the model output is a weighted sum of the attribute values that predicts the pair frequency. The learning process aims to obtain a regression model for the weights from the output of the data pre-processing step.

Table 1 reports different models learned using all the available pairs: for each model, the table reports the used features with their learned weights, the regression metrics provided by Weka, and the metric values for the recommendation of cold items (Fig. 1). Model-1 uses all the available features; Model-2 leaves out the index-based similarity about textual metadata; Model-3 leaves out the features based on recording and publication ages; Model-4 leaves out the

---

[3] http://www.cs.waikato.ac.nz/ml/weka/

Table 1 Learned models

	Model-1	Model-2	Model-3	Model-4	Model-5	Model-6
sameLang	x 1.2479	x 0.8097	x 1.0349	x 3.2801	x 1.7958	x 1.0363
sameDetectedLang	x -0.3375	x -0.2759	x	x	x	x -0.3374
freqLang	x 2.5217	x 4.6866	3.4824	x -3.8682	x	x 3.9266
freqDetectedLang	x -0.3458	x -0.2765	x -0.4943	x -0.2544	x -0.6531	x -0.3456
description	x 17.9226		x 16.5921	x 22.2335	x 20.9256	x 17.4239
descriptionLen	x -0.9129	x 1.3709	x -1.4296	x -1.0610	x -1.3196	x -0.9225
deltaPubAge	x	x -0.0557		x 0.0173	x -0.0963	x -0.0255
deltaRecAge	x -0.0856	x -0.0907		x -0.0776	x -0.0891	x -0.0828
pubAgeOlder	x 0.0890	x 0.1568		x -0.0194	x 0.1543	x 0.1252
pubAgeNewer	x -0.0417	x -0.0861		x	x -0.1626	x -0.0603
recAgeOlder	x 0.0692	x 0.0709		x 0.0401	x 0.0779	x 0.0730
recAgeNewer	x 0.0282	x 0.0111		x 0.0729	x 0.0512	x 0.0388
sameConference	x 4.4207	x 5.0438	x 4.3148		x 4.5780	x 4.5366
similarConference	x 3.1643	x 3.2060	x 3.3740		x 3.4212	x 2.9982
atLeastOneSharedAuth	x 3.8986	x 4.0301	x 1.7389	x 4.2799		x 3.4690
sharedAuth	x	x 1.6587	x 4.2296	x		x 0.7031
sharedAffil	x 3.2381	x	x 2.5147	x 4.6232		x 3.0751
categoryBest	x -2.9007	x -2.9065	x -3.6669	x -3.1974	x -3.0285	
categoryAvg	x 2.5258	x 2.0846	x 3.5300	x 0.6944	x 2.1317	
Correlation coefficient	0.1796	0.1736	0.1661	0.1579	0.1739	0.1719
Mean absolute error	5.9355	5.9498	5.8786	5.9772	5.9521	5.8783
Root mean squared error	23.2987	23.3244	23.3549	23.3868	23.323	23.3315
Relative absolute error	96.5899	96.8226	95.6633	97.2684	96.8598	95.6583
Root relative squared error	98.3737	98.482	98.6106	98.7453	98.4762	98.5119
MARp (preliminary)	0.06220	0.05752	0.05535	0.05990	0.01295	0.06051
MARp (final)	0.05492	0.04715	0.05306	0.05145	0.01163	0.05295

conferences; Model-5 leaves out the authors; Model-6 leaves out the categories. Some weights are missing for the fitness of the learning method.

The learned weights of a model are stored in a configuration file, with the option to add a boost factor for each weight to easily explore the feature influences beside the learned model. Fig. 2, Fig. 3 and Fig. 4 report the values of the evaluation metric for the recommendations using Model-1 when a boost factor is changed. The boost factors can be modified also to implement a naive feedback control on recommendations without performing a complete learning step. Please note that the learning step aims to obtain the best regression model to predict view frequencies for hot and cold item pairs starting from the training set with only hot items, while the recommendation evaluation concerns with list of cold items and the Challeng $MARp$ cannot be immediately used in the learning step.

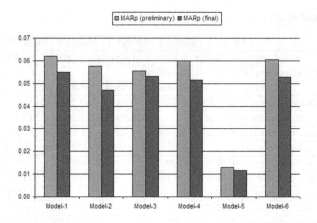

Fig. 1 Recommendation performances of models in Table 1

Fig. 5 reports the evaluation metric values for the submitted solutions when the boost factors for the learned weight in Model-1 are changed. Indeed, the Challenge execution contemplates a limited number of submissions to allow the parameter tuning, that is quite narrow compared to the probes reported in Fig. 2, Fig. 3 or Fig. 4. Anyway, the submitted solutions always outperform the random baseline (*MARp*: 0.01949).

Recommendation step   uses the in-memory representation of the pre-processing step and the learned weights to predict frequencies of lecture pairs. Each pair is composed of an *old* lecture from the task query in the training set and a *cold* lecture in the test set. The highest values are used to select the 30 cold items for the recommendation.

The in-memory representation and the lightweight prediction model allow to formulate a new recommendation in a reasonably short time.

The in-memory representation of the data pre-processing step is also used to create R^4 scripts to visualize the information in the dataset for an informed selection of the content-bases features. For instance, Fig. 6 shows how the views are temporally distributed considering the recording and publishing ages: the behavior is quite dissimilar for the two time scales, indeed, the oldest recorded lectures are seldom viewed as the cumulative box-plot and density function (the rightmost subgraphs) highlight, conversely the oldest published lectures have the highest density of views. Probably, the user interest for old lectures is weak even if the VideoLectures.Net kindled a lot of attention during the first months. In addition the views of lectures decrease when their recording and publishing ages decrease. Thus recent lectures need some assistance. Fig. 6 supports the idea to exploit age-based features in the model learning. In addition, Section 5.3 presents an initial investigation to filter lecture pairs for the learning step on the publishing time criterion. Fig. 7 shows how the views

---

[4] http://www.r-project.org/

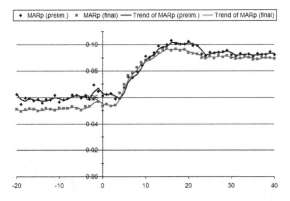

Fig. 2 Recommendation performances changing the boost factor of "categoryBest"

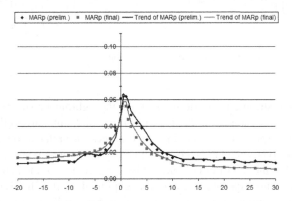

Fig. 3 Recommendation performances changing the boost factor of "deltaRecAge"

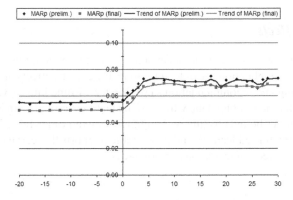

Fig. 4 Recommendation performances changing the boost factor of "similarConference"

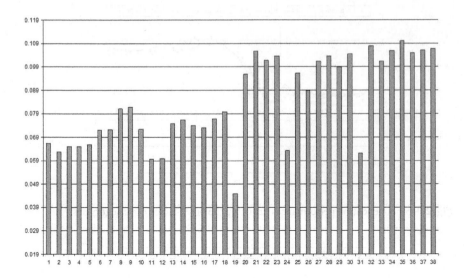

Fig. 5 Mean Average R-precision of submitted solutions

of each item are distributed considering its type: the rightmost histogram shows the cumulative views for each type; the uppermost box-plot summarizes the views for each items. Fig. 7 spots how the coldness and hotness are related to the item type. Fig. 8 shows how types and languages are linked by training pairs: the circular areas are proportional to the logarithm of cumulative frequencies for the pairs of lectures viewed together. This kind of information is exploited by the "freqLang" feature. For instance, the training pairs allows to state than pairs of lectures both in English are more probable than pairs of lectures in Dutch and English or pairs of lectures both in Dutch.

5.3   Time Effects

In Section 5.2 we witness the time dependency of lecture views, thus we filter the training set by the publication age of lectures to investigate the effects of such dependency on learning and recommendation steps. We try 14 time limits (with a step of 3 months) to filter the training data with the same features used to learn Model-1 but with different upper bounds for the pair common viewing time: Table 2 and Fig. 9 report the amount of used lectures and pairs, Table 3 reports the learned models and Fig. 10 reports the recommendation performances.

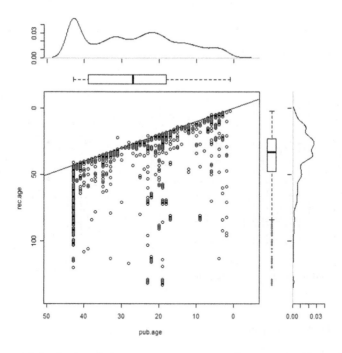

Fig. 6 Temporal distribution of views

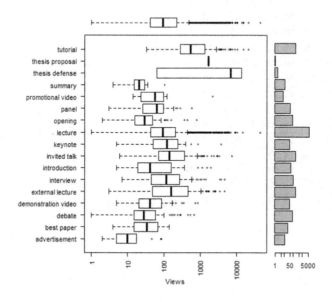

Fig. 7 Distribution of views considering item types

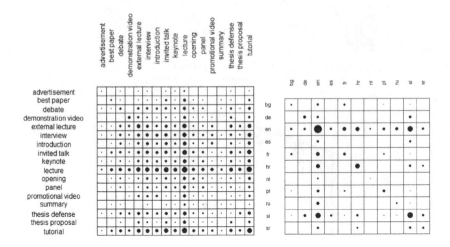

Fig. 8 Types and languages in lecture pairs

Table 2 Used lectures and pairs

Time limit	Lectures		Pairs	
none	6,936	100.0%	355,353	100.0%
42	5,506	79.4%	186,921	52.6%
39	5,269	76.0%	178,414	50.2%
36	4,857	70.0%	121,976	34.3%
33	4,555	65.7%	103,650	29.2%
30	3,912	56.4%	82,805	23.3%
27	3,476	50.1%	72,829	20.5%
24	3,298	47.5%	65,117	18.3%
21	2,137	30.8%	27,449	7.7%
18	1,756	25.3%	15,766	4.4%
15	1,420	20.5%	11,024	3.1%
12	976	14.1%	6,072	1.7%
9	707	10.2%	3,758	1.1%
6	474	6.8%	2,047	0.6%
3	187	2.7%	442	0.1%

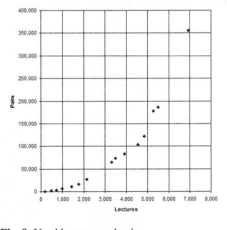

Fig. 9 Used lectures and pairs

Table 3 Models obtained from the same data used for Model-1 by changing the upper bound of pair common viewing time

Time limit	none	42	39	36	33	30	27	24	21	18	15	12	9	6	3
sameLang	1.248	2.504	2.567	2.921	3.017	3.540	3.551	1.322	-3.066	-5.034		-9.736			
sameDetectedLang	-0.338	-0.717	-0.731	-0.795	-0.910	-1.147	-1.166	-1.331	-0.947	1.126	1.658	2.318			
freqLang	2.522			-2.348	-2.412	-2.915	-2.400		17.470	21.128	20.128	39.254	2.430		
freqDetectedLang	-0.346	-0.720	-0.728	-0.793	-0.910	-1.149	-1.161	-1.330		1.126	1.657	2.318			
description	17.923	15.296	15.278	10.612	8.943	7.467	4.920	3.401	-12.263		12.623	11.015	3.248	5.392	-5.393
descriptionLen	-0.913	0.899	0.930	3.441	4.402	5.708	6.411	7.852	4.916	-4.790	-7.636	-9.418	-0.820	-1.171	
deltaPubAge		-0.080		0.260	0.250	0.703	0.071	0.044	0.467	-0.210		-0.728			
deltaRecAge	-0.086	0.133	-0.238	-0.221	-0.155	-0.049	-0.017	-0.454	-0.075	-0.045		0.091	-0.014		-0.031
pubAgeOlder	0.089	0.213	0.149	-0.155	-0.109	-0.536	0.062	0.103		0.176		1.424		0.451	
pubAgeNewer	-0.042	-0.217	-0.139	0.049		0.443	-0.160	-0.173	0.492	0.299	0.881	1.898	0.417	0.414	
recAgeOlder	0.069	-0.169	0.203	0.191	0.128	0.028		0.439	0.012			-0.091	0.024	0.024	0.027
recAgeNewer	0.028	0.349	-0.020	0.010	0.081	0.188	0.219	-0.222	0.084	0.070	-0.165	-0.182	0.035		0.135
sameConference	4.421	8.182	8.337	8.805	9.868	10.920	11.887	13.099	7.941	3.321	3.697	4.017	2.027	2.569	7.213
similarConference	3.164	0.500	0.497	-0.777	-1.037	-0.853	-1.306	-1.809	3.883	1.061		3.191			-6.713
atLeastOneSharedAuth	3.899	4.689	5.132	1.385	0.995	0.932	1.084	0.846	-0.840	3.224	1.952	-3.191	-1.984	-1.677	-4.063
sharedAuth		-3.811	-4.429	-0.946	-0.869	-1.771	-1.971	-1.916		-4.543		3.666	3.314	2.969	4.641
sharedAffil	3.238	5.220	5.373	3.753	3.208	2.611	2.325	1.698		-1.770	-9.038	-4.539	-1.564	-0.974	-1.443
categoryBest	-2.901	-1.739	-1.747	-1.365	-1.182	-1.718	-1.493	-2.207	-6.418	-8.205	-12.464	-18.930	-0.961		
categoryAvg	2.526	3.731	3.877	1.367	1.161	-1.209	-1.330	-2.427	-3.152	-17.577	-21.026	-13.618	2.674	4.373	1.525
Correlation coefficient	0.180	0.272	0.273	0.323	0.330	0.341	0.345	0.348	0.300	0.302	0.280	0.368	0.373	0.257	0.533
Mean absolute error	5.936	5.889	6.032	5.968	6.419	7.322	7.858	8.460	9.652	6.903	6.411	9.023	2.702	2.775	0.939
Root mean squared error	23.299	20.041	20.439	19.332	20.417	22.027	23.056	24.225	27.669	20.426	19.087	23.782	5.940	6.735	1.989
Relative absolute error	96.590	95.564	95.795	94.722	95.220	96.489	96.873	97.544	95.283	93.306	106.305	119.144	84.190	96.804	100.341
Root relative squared error	98.374	96.220	96.197	94.652	94.399	94.007	93.858	93.766	95.394	95.324	95.989	92.968	92.804	96.646	84.611
MARp (preliminary)	0.0622	0.0489	0.0455	0.0262	0.0217	0.0203	0.0187	0.0182	0.0380	0.0519	0.0588	0.0281	0.0081	0.0102	0.0054
MARp (final)	0.0549	0.0346	0.0312	0.0231	0.0190	0.0178	0.0180	0.0187	0.0404	0.0468	0.0527	0.0280	0.0102	0.0144	0.0050

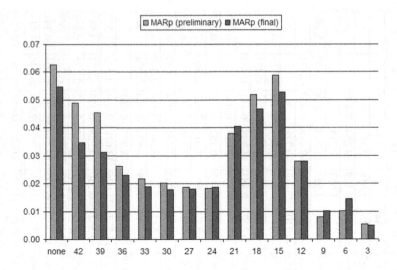

Fig. 10 Recommendation performances changing the upper bound of pair common viewing time

The recommendation performances appear to be laid in three regions. In the first (until the time limit of 24 months) performance worsen regularly beyond the baseline: this proceeding is legitimated by the progressive loss of data for the learning step. In the second region (for the time limits form 21 to 15 months), the performances increase. In the third region (for the time limits less then 12 months), the performance are the worst: indeed, the used lectures and pairs are really few because this region is between the moments t_1 and t_2 used by organizers to split the dataset (Section 3).

The second region is quite amazing because the performances are rather similar to ones obtained with the whole dataset. Probably, this region is the least noisy because it contains the newest *hot* lecture nearest to t_1 and, consequently, to the lectures in the test set.

Fig. 11, Fig. 12 and Fig. 13 report the feature influences on the recommendation performances using the same features used for the Model-1 in Fig. 2, Fig. 3 and Fig. 4. Two time limits (18 months (a) in the second region and 24 months (b) in the first region) are considered. The behaviour in the second region appears quite similar to the behaviour of Model-1: Fig. 11(a) and Fig. 2 have a similar shape, as well as Fig. 12(a) and Fig. 3, and Fig. 13(a) and Fig. 4. On the other hand, the behaviour at 24 months appears quite dissimilar to the behaviour of Model-1, even though the two models are in the same region.

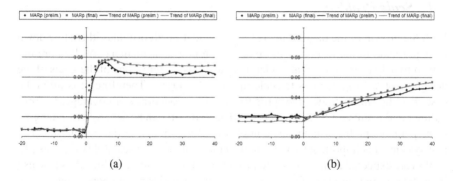

Fig. 11 Recommendation performances changing the boost factor of "categoryBest" with a upper bound of pair common viewing time of 18 months (a) and 24 months (b)

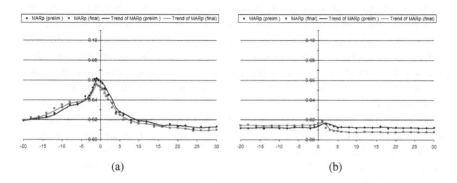

Fig. 12 Recommendation performances changing the boost factor of "deltaRecAge" with a upper bound of pair common viewing time of 18 months (a) and 24 months (b)

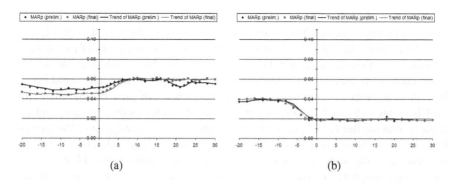

Fig. 13 Recommendation performances changing the boost factor of "similarConference" with a upper bound of pair common viewing time of 18 months (a) and 24 months (b)

5.4   Scale Problem

With the growth of the dataset, many recommendation algorithms either slow down or require additional resources as computation power or memory. As RSs are designed to help users to navigate in large collections of items, one of the goals of the designers of such systems is to scale up to real datasets. Therefore, it is often the case that algorithms trade other properties, such as accuracy or coverage, for providing rapid results for huge datasets [3]. The trade-off can be achieved by changing some parameters, for instance the complexity of the model, or the sample size. For real systems it is important to measure the compromises that scalability dictates [9].

RSs are expected in many cases to provide recommendation on-line, thus it is also important to measure how fast does the system provides recommendations [4, 7]. Common measurements are the number of recommendations that the system can provide per second (the throughput of the system) and the required time for making a recommendation (the latency or response time).

The developed Java components allow to complete the recommendation task for the 5,704 lectures in almost 85 seconds on a notebook with an Intel Core 2 at 2.0 GHz as CPU and 2GB of RAM, i.e., each new recommendation about 30 cold items over the selected 1,122 ones is provided in almost 15 milliseconds. Reasonably, a production server allows to reduce further the response time for new recommendations and a cache specifically devised for the recommendations allows to increase the throughput.

Moreover, the learning step performed by Weka absorbs the most resources. Although the step is designed to be performed off-line, the time and space requirements can be reduced by exploiting few features or less previous data. The six models presented in Section 5.2 and time effects discussed in Section 5.3 provides some hints in this direction.

6   Conclusions

We have described the steps to achieve the submitted solution that outperforms the random baseline. The in vitro evaluation of a solution to the cold start problem is an arduous task, since the common assumption about the reliability of past data to provide predictions is weakened. For instance, Fig. 14 shows how many of the old items used in the evaluation of submitted solutions have few associated cold items. Fig. 14 is obtained by exploring the actual frequencies of view of cold items for the Challenge task. Such frequencies are used to obtain the evaluation score and are made available after the Challenge to encourage further investigations. The lack of links between old items and cold items comes from the real data and it warrants the need for some strategy to deal with cold items. In additions, Fig. 14 shows that the average frequency of the considered pairs of old and cold lectures increases when the users view an increasing number of cold items for the same old item: the transition from cold to hot seems to be on the highest levels used for the evaluation

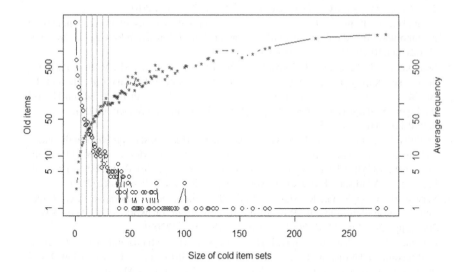

Fig. 14 Number of old items (o) and their pair average frequencies (*) on the size of cold item sets used in the evaluation

metric. The evaluation cut-off lengths for *MARp* (5, 10, 15, 20, 25, 30) are shown in Fig. 14 as grey vertical lines.

The idea of integrating a content-based approach allows to provide also serendipitous recommendations alongside classical ones [5]. Indeed the content-based item similarity can be used to obtain a hybrid RS that exploits the "Anomalies and exceptions" approach [11] to spot potential serendipitous items as further trade-off with the entire system accuracy.

Finally, the scalability performance is considered as a primary requirement and a lightweight solution is pursued. Beside other considerations, providing recommendations about cold items remains a chancy task, thus a computational resource curtailment for such task is a reasonable strategy to control performance trade-off of a RS. The preliminary performance for the notebook execution is quite promising and some future directions for improving latency and throughput are sketched.

References

1. Antonov-Fantulin, N., Bošnjak, M., Žnidaršič, M., Grčar, M., Morzy, M., Šmuc, T.: ECML-PKDD 2011 Discovery Challenge Overview. In: Šmuc, T., Antonov-Fantulin, N., Morzy, M. (eds.) Proceedings of the ECML/PKDD Discovery Challenge Workshop, CEUR Workshop Proceedings, vol. 770, pp. 7–20. CEUR-WS.org (2011)
2. Burke, R.: Hybrid recommender systems: Survey and experiments. User Modeling and User-Adapted Interaction 12, 331–370 (2002), doi:10.1023/A:1021240730564

3. Das, A.S., Datar, M., Garg, A., Rajaram, S.: Google news personalization: scalable on-line collaborative filtering. In: Proc. of the 16th Int. Conf. on World Wide Web (WWW 2007), pp. 271–280. ACM (2007), doi:10.1145/1242572.1242610

4. Herlocker, J., Konstan, J.A., Riedl, J.: An empirical analysis of design choices in neighborhood-based collaborative filtering algorithms. Information Retrieval 5, 287–310 (2002), doi:10.1023/A:1020443909834

5. Iaquinta, L., de Gemmis, M., Lops, P., Semeraro, G., Filannino, M., Molino, P.: Intro-ducing serendipity in a content-based recommender system. In: Xhafa, F., Herrera, F., Abraham, A., Köppen, M., Bénitez, J.M. (eds.) Proc. of the 8th Int. Conf. on Hybrid Intelligent Systems (HIS 2008), pp. 168–173. IEEE Computer Society (2008), doi:10.1109/HIS.2008.25

6. Iaquinta, L., Semeraro, G.: Lightweight approach to the cold start problem in the video lecture recommendation. In: Šmuc, T., Antonov-Fantulin, N., Morzy, M. (eds.) Proceed-ings of the ECML/PKDD Discovery Challenge Workshop, CEUR Workshop Proceed-ings. CEUR Workshop Proceedings, vol. 770, pp. 83–94. CEUR-WS.org (2011)

7. Sarwar, B., Karypis, G., Konstan, J., Reidl, J.: Item-based collaborative filtering rec-ommendation algorithms. In: Proc. of the 10th Int. Conf. on World Wide Web (WWW 2001), pp. 285–295. ACM (2001), doi:10.1145/371920.372071

8. Schein, A.I., Popescul, A., Ungar, L.H., Pennock, D.M.: Methods and metrics for cold-start recommendations. In: Proc. of the 25th ACM SIGIR Conf. on Research and Devel-opment in Information Retrieval (SIGIR 2002), pp. 253–260. ACM (2002), doi:10.1145/564376.564421

9. Shani, G., Gunawardana, A.: Evaluating recommendation systems. In: Ricci, F., Rokach, L., Shapira, B., Kantor, P.B. (eds.) Recommender Systems Handbook, pp. 257–297. Springer (2011), doi: 10.1007/978-0-387-85820-3_8

10. Shardanand, U., Maes, P.: Social Information Filtering: Algorithms for Automating "Word of Mouth". In: Proceedings of ACM CHI 1995 Conference on Human Factors in Computing Systems, pp. 210–217 (1995), doi:10.1145/223904.223931

11. Toms, E.G.: Serendipitous information retrieval. In: DELOS Workshop: Information Seeking, Searching and Querying in Digital Libraries (2000)

Content-Based Keywords Extraction
and Automatic Advertisement Associations
to Multimodal News Aggregations

Giuliano Armano, Alessandro Giuliani, Alberto Messina, Maurizio Montagnuolo, and Eloisa Vargiu

Abstract. Nowadays, Web is characterized by a growing availability of multimedia data together with a strong need for integrating different media and modalities of interaction. Hence, one of the main challenges is to bring into the Web data thought and produced for different media, such as TV or press content. In this scenario, we focus on multimodal news aggregation retrieval and fusion. Multimodality, here, is intended as the capability of processing, gathering, manipulating, and organizing data from multiple media. In particular, we tackle two main issues: to extract relevant keywords to news and news aggregations, and to automatically associate suitable advertisements to aggregated data. To achieve the first goal, we propose a solution based on the adoption of extraction-based text summarization techniques; whereas to achieve the second goal, we developed a contextual advertising system that works on multimodal aggregated data. To assess the proposed solutions, we performed experiments on Italian news aggregations. Results show that, in both cases, the proposed solution performs better than the adopted baseline solutions.

1 Introduction

Modern broadcasters are facing an unprecedented technological revolution from traditional dedicated equipment to commodity hardware and software components, and from yesterday-one-to-many delivery paradigms to nowadays-Internet-based interactive platforms. Thus, information engineering and integration plays a vital role in

G. Armano · A. Giuliani · E. Vargiu
University of Cagliari, Dept.of Electrical and Electronic Engineering, Piazza d'Armi, I09123 Cagliari, Italy
e-mail: {armano,alessandro.giuliani,vargiu}@diee.unica.it

A. Messina · M. Montagnuolo
RAI Centre for Research and Technological Innovation, C.so Giambone, 68, I10135 Torino, Italy
e-mail: {a.messina,maurizio.montagnuolo}@rai.it

C. Lai et al. (Eds.): New Challenges in Distributed Inf. Filtering and Retrieval, SCI 439, pp. 33–52.
springerlink.com   © Springer-Verlag Berlin Heidelberg 2013

optimizing costs and quality of the provided services, and in reducing the "time to market" of data.

In this challenging scenario, this chapter focuses on multimodal news aggregation, retrieval, and fruition. Multimodality is intended as the capability of processing, gathering, manipulating, and organizing data from multiple media (e.g., television, press, the Internet) and made of different modalities such as audio, speech, text, image, and video. In our work, news come from two inputs: (i) automatically extracted, chaptered and transcribed TV news, and (ii) RSS feeds from online newspapers and press agencies.

The aim of this chapter is twofold: to extract keywords from news and news aggregations in order to concisely describe their content, and to automatically associate suitable advertisements to aggregated data. As for the former, we present a study aimed at automatically generating tag clouds for representing the content of multimodal aggregations (MMAs) of news information from television and from the Internet. To this end, we propose a solution based on Text Summarization (TS) and we make experiments to compare classical extraction-based TS techniques with respect to a simple technique based on part-of-speech (POS) tagging. As for the latter, we propose a system for contextually associating ads with news stories aggregations. In accordance to state-of-the-art approaches on Contextual Advertising (CA), the system adopts both syntactical and semantic techniques. Syntax is exploited by adopting a suitable TS technique aimed at reducing the size of the data while preserving the original meaning. Semantics is exploited by using a centroid-based classifier devoted to capture the main topics concerning with a given news aggregation.

This chapter extends and revises our previous work [2], the main extension being the proposal of the CA system.

The rest of the chapter is organized as follows. Section 2 recalls relevant work on information fusion, heterogeneous data clustering, TS, and CA. Section 3 sketches a reference scenario aimed at highlighting the usefulness of the proposed solutions. In Section 4, we recall the model for multimodal aggregation previously presented in [26] and we illustrate how news are stored according to that model. Section 5 focuses on the first problem addressed in this chapter by describing the adopted extraction-based TS techniques. Section 6 focuses on the second problem addressed in this chapter presenting our solution aimed at applying the proposed approach in a CA system. In Section 7, we illustrate our experiments aimed at assessing the proposed solutions. In Section 8, we show how the proposed solutions work in the selected reference scenario. Section 9 ends the chapter reporting conclusions.

2 Background

2.1 *Information Fusion and Heterogeneous Data Clustering*

Information (or data) fusion can be defined as the set of methods that combine data from multiple sources and use the obtained information to discover additional knowledge, potentially not discoverable by the analysis of the individual sources.

First attempts to organize a theory have been done in [31], in which the author proposes a cross-document structure theory and a taxonomy of cross-document relationships. Recently, some proposals have been made to provide a unifying view. The work in [15] classifies information fusion systems in terms of the underlying theory and formal languages. Moreover, in [23], the author describes a method (Finite Set Statistics) which unifies most of the research on information fusion under a Bayesian paradigm.

Many information fusion approaches currently exist in many areas of research, e.g., multi-sensor information fusion, notably related to military and security applications, and multimedia information fusion. In the latter branch, the closest to the present research, the work in [36] analyses best practices for selection and optimization of multimodal features for semantic information extraction from multimedia data. More recent relevant works are [30] and [19]. In [30], the authors present a self-organizing network model for the fusion of multimedia information. In [19], the authors implement and evaluate a fusion platform implementing the framework within a recommendation system for smart television in which TV programme descriptions coming from different sources of information are fused.

Heterogeneous data clustering is the usage of techniques and methods to aggregate data objects that are different in nature, for example video clips and textual documents. A type of heterogeneous data clustering is co-clustering, which allows simultaneous clustering of the rows and columns of a matrix. Given a set of m rows in n columns, a co-clustering algorithm generates co-clusters, i.e., a subset of rows which exhibit similar behavior across a subset of columns, or vice-versa. One of the first methods conceived to solve the co-clustering of documents using word sets as features is represented by [20], where RSS items are aggregated according to a taxonomy of topics. More challenging approaches are those employing both cross-modal information channels, such as radio, TV, the Internet, and multimedia data [13, 37].

2.2 Text Summarization

Automatic text summarization is a technique in which a text is summarized by a computer program. Given a text, a summarized text, which is a non redundant extract from the original text, is returned.

Radev et al. [32] define a summary as "a text that is produced from one or more texts, that conveys important information in the original text(s), and that is no longer than half of the original text(s) and usually significantly less than that". This simple definition highlights three important aspects that characterize the research on automatic summarization: (i) summaries may be produced from a single document or multiple documents; (ii) summaries should preserve important information; and (iii) summaries should be short. Unfortunately, attempts to provide a more elaborate definition for this task are in disagreement within the community [12].

Let us also recall the distinctions among different kinds of summaries given by Mani [24]: an *extract* consists entirely of material copied from the input; an *abstract* contains material that is not present in the input, or at least expresses it in a different way; an *indicative abstract* is aimed at providing a basis for selecting documents for closer study of the full text; an *informative abstract* covers all the salient information in the source at some level of detail; and *critical abstract* evaluates the subject matter of the source document, expressing the abstractor views on the quality of the author's work.

Summarization techniques can be divided in two groups [17]: those that extract information from the source documents (*extraction-based approaches*) and those that abstract from the source documents (*abstraction-based approaches*). The former impose the constraint that a summary uses only components extracted from the source document. These approaches put strong emphasis on the form, aiming to produce a grammatical summary, which usually requires advanced language generation techniques. The latter relax the constraints on how the summary is created. These approaches are mainly concerned with what the summary content should be, usually relying solely on extraction of sentences.

Although potentially more powerful, abstraction-based approaches have been far less popular than their extraction-based counterparts, mainly because generating the latter is easier. While focusing on information retrieval, one can also consider topic-driven summarization, which assumes that the summary content depends on the preferences of the user and can be assessed via a query, making the final summary focused on a particular topic. Since in this chapter we are interested in extracting suitable keywords, we exclusively focus on extraction-based methods.

An extraction-based summary consists of a subset of words from the original document and its bag of words (*BoW*) representation can be created by selectively removing a number of features from the original term set. Typically, an extraction-based summary whose length is only 10-15% of the original is likely to lead to a significant feature reduction as well. Many studies suggest that even simple summaries are quite effective in carrying over the relevant information about a document. From a text categorization perspective, their advantage over specialized feature selection methods lies in their reliance on a single document (the one that is being summarized) without computing the statistics for all documents sharing the same category label, or even for all documents in a collection. Moreover, various forms of summaries become ubiquitous on the Web and in certain cases their accessibility may grow faster than that of full documents.

Earliest instances of research on summarization of scientific documents extract salient sentences from text using features like word and phrase frequency [22], position in the text [6], and key phrases [14]. Various works published since then had concentrated on other domains, mostly on newswire data [29]. Many approaches addressed the problem by building systems dependent on the type of the required summary.

2.3   Contextual Advertising

Online Advertising is an emerging research field, at the intersection of Information Retrieval, Machine Learning, Optimization, and Microeconomics. Its main goal is to choose the right ads to present to a user engaged in a given task, such as sponsored search advertising or contextual advertising. Sponsored search advertising (or paid search advertising) puts ads on the page returned from a Web search engine following a query. CA (or content match) puts ads within the content of a generic, third party, Web page. A commercial intermediary, the ad network, is usually in charge of optimizing the selection of ads with the twofold goal of increasing revenue (shared between publisher and ad network) and improving user experience. In other words, CA is a form of targeted advertising for ads appearing on Web sites or other media, such as content displayed in mobile browsers. The ads themselves are selected and served by automated systems based on the content displayed to the user.

A natural extension of search advertising consists of extracting phrases from the target page and matching them with the bid phrases of ads. Yih et al. [38] proposed a system for phrase extraction, which uses a variety of features to determine the importance of page phrases for advertising purposes. To this end, the authors proposed a supervised approach that relies on a training set built using a corpus of pages in which relevant phrases have been annotated by hand.

Ribeiro-Neto et al. [33] examined a number of strategies to match pages and ads based on extracted keywords. They represented both pages and ads in a vector space and proposed several strategies to improve the matching process. In particular, the authors explored the use of different sections of ads as a basis for the vector, mapping both page and ads in the same space. Since there is a discrepancy between the vocabulary used in the pages and in the ads (the so called *impedance mismatch*), the authors improved the matching precision by expanding the page vocabulary with terms from similar pages.

In a subsequent work, Lacerda et al. [18] proposed a method to learn the impact of individual features by using genetic programming. The results showed that genetic programming helps to find improved matching functions.

Broder et al. [8] classified both pages and ads according to a given taxonomy and matched ads to the page falling into the same node of the taxonomy. Each node is built as a set of bid phrases or queries corresponding to a certain topic. Results showed a better accuracy than that corresponding to the classic systems (i.e., systems based on syntactic matching only). Let us also note that, to improve performances, this system may be used in conjunction with more general approaches.

Another approach that combines syntax and semantics has been proposed in [4]. The corresponding system, called ConCA (Concepts on Contextual Advertising), relies on ConceptNet, a semantic network able to supply commonsense knowledge [21].

Nowadays, ad networks need to deal in real time with a large amount of data, involving billions of pages and ads. Hence, efficiency and computational costs are crucial factors in the choice of methods and algorithms. Anagnostopoulos et al. [1] presented a methodology for Web advertising in real time, focusing on the contributions of the different fragments of a Web page. This methodology allows to identify short but informative excerpts of the Web page by means of several text summarization techniques, used in conjunction with the model developed in [8].

According to this view, Armano et al. [3, 5] studied the impact of the syntactic phase on CA. In particular, they performed a comparative study on text summarization in CA, showing that effective text summarization techniques may help to improve the behavior of a CA system.

Since bid phrases are basically search queries, another relevant approach is to view CA as a problem of query expansion and rewriting. Murdock et al. [28] considered a statistical machine translation model to overcome the problem of the impedance mismatch between pages and ads. To this end, they proposed and developed a system able to re-rank the ad candidates based on a noisy-channel model. In a subsequent work, Ciaramita et al. [10] used a machine learning approach, based on the model described in [8], to define an innovative set of features able to extract the semantic correlations between the page and the ad vocabularies.

3   A Reference Scenario

Let us consider a family composed by Bob, Alice, and their son Edinson. Bob is an investment broker. Due to his job, he is interested in both economy/finance news and transportation services. Alice is a housewife who cares with the health of her child. Edinson is a sport lover and his favorite sport is football.

Each family member is encountering severe problems for fulfilling personal information needs. From the Internet point of view, the availability on a daily basis of a wide variety of information sources generates a disproportionately high amount of content (e.g., newspaper articles and news agency releases) that makes it impossible for each of them to read everything that is produced. Furthermore, it is obvious that the Internet is not (yet) the only source of news information, being traditional media based on television channels still far from being left out in the near future. Due to the heterogeneity of individual interests, classical newscast programmes and TV advertising can be extremely inefficient, leaving viewers annoyed and upset.

To accomplish the users' needs, media industry wish would be to have an application able to aggregate data produced from different sources, to give a short description of them in form of keywords, and to associate them with advertising messages according to the content of the generated aggregations. In this way the map is complete and each user can get informed with completeness and efficacy. For example, Bob might stay tuned on last stock market news through his tablet, while being advised on journeys and transports. Alice might watch healthcare news stories on her television, while being recommended on the healing properties of herbs and natural remedies. Finally, Edinson might browse his favorite team articles, while getting hints on new sport furniture. An example of real advertisements that would be suggested to Edinson is illustrated in Figure 7.

4 Multimodal Aggregation

Multimodal aggregation of heterogeneous data, also known as *information mash-up*, is a hot topic in the World Wide Web community. A multimodal aggregator is a system that merges content from different data sources (e.g., Web portals, IPTV, etc.) to produce new, hybrid data that was not originally provided. Here, the challenge lies in the ability of combining and presenting heterogeneous data coming from multiple information sources, i.e., *multimedia*, and consisting of multiple types of content, i.e., *cross-modal*. As a result of this technological breakthrough, the content of modern Web is characterized by an impressive growth of multimedia data, together with a strong trend towards integration of different media and modalities of interaction. The mainstream paradigm consists in *bringing into the Web* what was thought (and produced) for different media, like TV content (acquired and published on websites and then made available for indexing, tagging, and browsing). This gives rise to the so-called *Web Sink Effect*. This effect has rapidly started, recently, to unleash an ineluctable evolution from the original concept of the Web as a resource where to *publish* things produced in various forms *outside* the Web, to a world where things *are born and live* on the Web. In this chapter, we adopt Web newspaper articles and TV newscasts as information sources to produce multimodal aggregations of informative content integrating items coming from both contributions. In the following of this section we briefly overview the main ideas behind this task (the interested reader may refer to [26], for further details).

The corresponding system can be thought as a processing machine having two inputs, i.e., digitized broadcast news streams (DTV) and online newspapers feeds (RSSF), and one output, i.e., the multimodal aggregations that are automatically determined from the semantic aggregation of the input streams by applying a co-clustering algorithm whose kernel is an asymmetric relevance function between information items [27].

Television news items are automatically extracted from the daily programming of several national TV channels. The digital television stream is acquired and partitioned into single programmes. On such programmes, newscast detection and segmentation into elementary news stories is performed. The audio track of each story is finally transcribed by a speech-to-text engine and indexed for storage and retrieval. Further details can be found in [25].

The RSSF stream consists of RSS feeds from several major online newspapers and press agencies. Each published article is downloaded, analyzed, and indexed for search purposes. The first step of the procedure consists in cleaning the downloaded article Web pages from boilerplate content, i.e., HTML markups, links, scripts, and styles. Linguistic analysis, i.e., sentence boundary detection, sentence tokenization, word lemmatization, and POS tagging, is then performed on the extrapolated contents. The output of this analysis is then used to transform the RSS content into a query to access the audio transcriptions of the DTV news stories, thus allowing to combine text and multimedia in an easy way.

The output of the clustering process is a set of multimodal aggregations of broadcast news stories and newspaper articles related to the same topic. TV news stories and Web newspaper articles are fully cross-referenced and indexed. For each multimodal aggregation, users can use automatically extracted tag clouds, to perform local or Web searches. Local searches can be performed either on the specific aggregation the tags belong to or to the global set of discovered multimodal aggregations. Tag clouds are automatically extracted from each thread topic as follows: (i) each word classified as proper noun by the linguistic analysis is a tag; (ii) a tag belongs to a multimodal aggregation if it is present in at least one aggregated news article; and (iii) the size of a tag is proportional to the cumulative duration of television news items which are semantically relevant to the aggregated news article to which the tag belongs. In so doing, each news aggregation, also called *subject*, is described by a set of attributes, the main being:

- *info*, the general information included title and description;
- *categories*, the set of most relevant categories to which the news aggregation belong. They are automatically assigned by AI:Categorizer[1], trained with radio programme transcriptions, according to a set of journalistic categories (e.g., *Politics*, *Currents Affairs*, *Sports*);
- *tagclouds*, a set of automatically-generated keywords;
- *items*, the set of Web articles that compose the aggregation;
- *videonews*, the collection of relevant newscasts stories that compose the news aggregation.

Hence, a news aggregation is composed by online articles (*items*) and parts of newscasts (*videonews*). In this chapter, we concentrate only in the former. Each item is described as set of attributes, such as:

- *pubdate*, the timestamp of first publication;
- *lastupdate*, the timestamp when the item was updated;
- *link*, the URL of the news Web page;
- *feed*, the RSS feed link that includes the item;
- *title*, the title;
- *description*, the content;
- *category*, the category to which the news belong (according to the previously mentioned classification procedure);
- *keywords*, the keywords automatically extracted as described above.

5 Content-Based Keyword Extraction to Multimodal News Aggregations

The first aim of this work is to automatically extract keywords to news and news aggregations. In particular, we are aimed at selecting keywords relevant to the news and news aggregations.

Among other solutions, we decided to use suitable extraction-based TS techniques. To this end, we first consider six straightforward but effective extraction-based text summarization techniques proposed and compared in [17] (in all cases, a

---

[1] http://search.cpan.org/ kwilliams/AI-Categorizer-0.09/lib/AI/Categorizer.pm

word occurring at least three times in the body of a document is a keyword, while a word occurring at least once in the title of a document is a title-word):

- *Title* (T), the title of a document;
- *First Paragraph* (FP), the first paragraph of a document;
- *First Two Paragraphs* (F2P), the first two paragraphs of a document;
- *First and Last Paragraphs* (FLP), the first and the last paragraphs of a document;
- *Paragraph with most keywords* (MK), the paragraph that has the highest number of keywords;
- *Paragraph with Most Title-words* (MT), the paragraph that has the highest number of title-words.

Let us note that we decided to not consider the Best Sentence technique, i.e. the technique that takes into account sentences in the document that contain at least 3 title-words and at least 4 keywords. This method was defined to extract summaries from textual documents such as articles, scientific papers and books. In fact, news are often inadequate to find meaningful sentences composed by at least 3 title-words and 4 keywords in the same sentence.

Furthermore, we consider the enriched techniques proposed in [3]:

- *Title and First Paragraph* (TFP), the title of a document and its first paragraph:
- *Title and First Two Paragraphs* (TF2P), the title of a document and its first two paragraphs;
- *Title, First and Last Paragraphs* (TFLP), the title of a document and its first and last paragraphs;
- *Most Title-words and Keywords* (MTK), the paragraph with the highest number of title-words and that with the highest number of keywords.

One may argue that the above methods are too simple. However, as shown in [7], extraction-based summaries of news articles can be more informative than those resulting from more complex approaches. Also, headline-based article descriptors proved to be effective in determining user's interests [16]. Moreover, these techniques have been successfully applied in the contextual advertising field [5].

6   Automatic Advertisement Associations to Multimodal News Aggregations

The second aim of this work is to automatically suggest relevant advertisements to news and news aggregations. To this end, we developed a suitable contextual advertising system.

The proposed system aims to suggest ads that match with the content of a given news aggregation. To this end, we adopt a solution compliant with state-of-the-art CA approaches, such as those proposed in [1, 5, 8].

Figure 1 illustrates the main components of the prosed system: the *BoW and CF Extractor*, aimed at extracting the Bag-of-Words (*BoW*) and the Classification Features (*CF*) of a given news aggregation, and the *Matcher* aimed at selecting the ads according to the similarity with the given news aggregation.

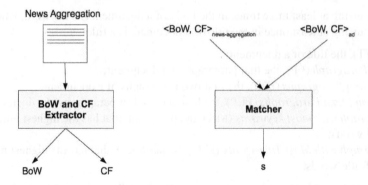

Fig. 1 The main components of the proposed system. For each news aggregation, both syntactic and semantic information are extracted. Syntactic information is expressed as a bag-of-words (BoW) vector. Semantic information is expressed as a weighted classification feature (CF) vector. The most prominent ads with respect to the news aggregation content are those whose similarity score *s* are above a given threshold.

Figure 2 shows the process of extracting syntactical and semantic features from a news aggregation performed by the *BoW* vector generator and the *CF* vector extractor, respectively. Given a news aggregation, the *News Extractor* is aimed at extracting all the news that compose it. In order to transform the news content into an easy-to-process document, any given news is also parsed to remove stop-words, tokenize it and stem each term. Then, for each news, the *Text Summarizer* calculates a vector representation as *BoW*, each word being weighted by TF-IDF [35]. According to the comparative experiments illustrated in the previous Section, we implemented the text summarization technique that showed the best results, i.e., TF2P. This extraction-based technique takes into account information belonging to the Title and the First Two Paragraphs of the news [3].

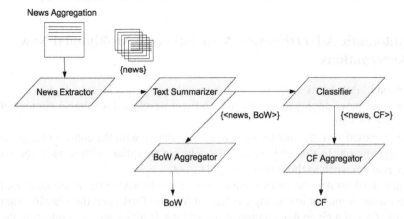

Fig. 2 The main modules involved in the process of BoW and CF extraction.

The output of the *Text Summarizer*, i.e., a list of ⟨*news*, *BoW*⟩ pairs, is given in input to the *BoW Aggregator* that is devoted to calculate the *BoW* of the whole news aggregation. The aggregated *BoW* is obtained considering the occurrences in the whole set of news, weighted by TF-IDF. Since, typically, CA systems work with a sole Web page this module is absent in classical CA systems, its goal being to allow us to work with aggregated data. To infer the topics of each news, the *Classifier* analyzes them according to a given set of classes based on a taxonomy of journalistic categories. First, for each class we represent it with its centroid, calculated starting from the training set. We then classify each document by adopting the Rocchio classifier [34] with only positive examples and no relevance feedback. Each centroid component is defined as a sum of TF-IDF values of each term, normalized by the number of documents in the class. The classification is based on the cosine of the angle between the news and the centroid of each class. The score is normalized by the news and class lengths to produce a comparable score. The output of this module is a list of ⟨*news*, *CF*⟩ pairs, where, in accordance with the work by Broder et al. [8], *CF* are the Classification Features extracted by the classifier. The output of the *Classifier*, i.e., a list of ⟨*news*, *CF*⟩ pairs, is given in input to the *CF Aggregator* that is devoted to calculate the *CF* of the whole news aggregation. The aggregated *CF* is obtained considering the scores giving by the classifier. It is worth noting that, similarly to the *BoW Aggregator*, this module, absent in classical CA systems, allows us to work with aggregated data.

Each ad, which in our work is represented by the Web page of a product or service company, is processed in a similar way and it is represented by suitable *BoW* and *CF*. To choose the ads relevant to a news aggregation, the *Matcher* assigns a score s to each ad according to its similarity with a given news:

$$s(n,a) = \alpha \cdot sim_{BoW}(n,a) + (1-\alpha) \cdot sim_{CF}(n,a) \tag{1}$$

in which α is a global parameter that permits to control the impact of *BoW* with respect to *CF*, whereas $sim_{BoW}(n,a)$ and $sim_{CF}(n,a)$ are cosine similarity scores between the news (n) and the ad (a) using *BoW* and *CF*, respectively. For $\alpha = 0$ only semantic analysis is considered, whereas for $\alpha = 1$ only the syntactic analysis is considered.

7   Experiments and Results

To assess the effectiveness of the proposed solutions, we perform several experiments. As for the task of extracting keywords, we performed two sets of comparative experiments: (i) experiments on the sole news comparing the performance with those corresponding to the adoption of the keywords provided in the *keyword* attribute and (ii) experiments on news aggregations comparing the performance with those corresponding to the adoption of the keywords provided in the *tagclouds* attribute. Performances have been calculated in terms of precision, recall, and F1 by exploiting a suitable classifier. As for the task of associating suitable ads, we, first, set up a dataset of ads, composed by Web pages of product-service companies, and, then, about 15 users[2] were asked to evaluate the relevance of the suggested ads. Performances have been calculated in terms of *precision at k*, i.e., the precision of the system in suggesting *k* ads, with *k* varying from 1 to 5.

7.1 The Adopted Dataset

As for the comparative study on TS, experiments have been performed on about 45,000 Italian news and 4,800 news aggregations from January 16, 2011 to May 26, 2011. The adopted dataset is composed by XML files, each one describing a subject according to the attributes described in Section 4. News and news aggregations were previously classified into 15 categories, i.e., the same categories adopted for describing news and news aggregations.

As for assessing the performances of the proposed CA system, experiments were performed on a set of about 600 news aggregations belonging to the following 8 categories[3]: *Economy and Finance* (EF), *Environment Nature and Territory* (ENT), *Health and Health Services* (HHS), *Music and Shows* (MS), *Publishing Printing and Mass Media* (PPMM), *Religious Culture* (RC), *Sports* (S), and *Transportation* (T). For each category we selected 11 ads as Web pages of product-service companies.

7.2 Experimenting Text Summarization

First, we performed experiments on news by adopting a system that takes as input an XML file that contains all the information regarding a news aggregation. For each TS technique, first the system extracts the news, parses each of them, and adopts stop-word removing and stemming. Then, it applies the selected TS technique to extract the corresponding keywords in a vector representation (*BoW*). To calculate the effectiveness of that technique, the extracted *BoW* is given as input to a centroid-based classifier, which represents each category with a centroid calculated starting from a suitable training set[4]. A *BoW* vector is then classified by measuring the distance between it and each centroid, by adopting the cosine similarity measure.

Performances are calculated in terms of precision, recall, and F1. As for the baseline technique (B), we considered the BoW corresponding to the set of keywords of the *keywords* attribute. Table 1 summarizes the results.

Subsequently, we performed experiments on news aggregations in a way similar to the one adopted for the sole news. For each TS technique, first the system

---

[2] Assessors have been selected among students and young researchers of the Department of Electrical and Electronic Engineering as well as workers at RAI Centre of Research and Technological Innovation.

[3] We do not take into considerations categories with a very few numbers of documents and also those that, according to [9], are not suitable to suggest ads.

[4] In order to evaluate the effectiveness of the classifier, we performed a preliminary experiment in which news are classified without using TS. The classifier shown a precision of 0.862 and a recall of 0.858.

Table 1 Comparisons among TS techniques on news.

	B	**T**	**FP**	**F2P**	**FLP**	**MK**	**MT**	**TFP**	**TF2P**	**TFLP**	**MTK**
P	0.485	0.545	0.625	0.705	0.681	0.669	0.650	0.679	**0.717**	0.706	0.692
R	0.478	0.541	0.594	0.693	0.667	0.665	0.640	0.663	**0.704**	0.697	0.686
F1	0.481	0.543	0.609	0.699	0.674	0.667	0.645	0.671	**0.710**	0.701	0.689
# terms	5	5	13	23	22	20	15	16	25	24	18

processes each news belonging to the news aggregation in order to parse it, to disregard stop-words, and to stem each remaining term. Then, it applies to each news the selected TS technique in order to extract the corresponding keywords in a *BoW* representation. Each extracted *BoW* is then given in input to the same centroid-based classifier used for the news. The category to which the news aggregation belongs to is then calculated averaging the scores given by the classifier for each involved item.

Table 2 shows the results obtained by comparing each TS technique, the baseline (B) being the BoW corresponding to the set of keywords of the *tagclouds* attribute.

Table 2 Comparisons among TS techniques on news aggregations.

	B	**T**	**FP**	**F2P**	**FLP**	**MK**	**MT**	**TFP**	**TF2P**	**TFLP**	**MTK**
P	0.624	0.681	0.693	0.764	0.734	0.717	0.727	0.731	**0.770**	0.766	0.737
R	0.587	0.678	0.683	0.766	0.728	0.709	0.718	0.728	**0.769**	0.759	0.729
F1	0.605	0.679	0.688	0.765	0.731	0.713	0.723	0.729	**0.769**	0.762	0.733
# terms	62	70	204	319	337	231	200	215	318	338	280

Results clearly show that, for both news and news aggregations, TS improves performances with respect to the adoption of the baseline keywords. In particular, best performances in terms of precision, recall, and –hence– F1, are obtained by adopting the TF2P technique. The last row of Table 1 and Table 2 shows the number of terms extracted by each TS technique. It is easy to note that, except for the T technique, TS techniques extract a number of terms greater than that extracted by the baseline approach. Let us also note that precision, recall, and F1 calculated for news aggregations are always better than those calculated for news. This is due to the fact that news aggregations are more informative than the sole news and the number of extracted keywords is greater.

7.3 Experimenting the Contextual Advertising System

About 15 users were asked to evaluate the relevance of the suggested advertisements and performance were calculated in terms of *precision at k*, i.e., the precision of the system in suggesting k ads, with k varying from 1 to 5.

Preliminary experiments have been performed to calculate the best value of α in Equation 1 to maximize the number of correct proposed advertisements. Figure

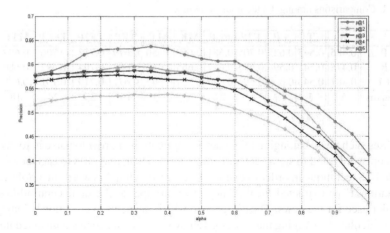

Fig. 3 *Precision at k*, varying α.

3 shows the results obtained comparing, for each suggested $\langle newsaggregation, ad \rangle$ pair, the category to the news aggregation belongs with the one to the ad belongs, varying α. Results show that the best results are obtained with a value of α in the range $0.25 - 0.40$, meaning that the impact given by the semantic contribution is greater.

Then, we set α to 0.35 and we asked to the selected assessors to give, for 80 randomly selected news aggregations, a degree of relevance, i.e., relevant (1), somewhat relevant (2), or irrelevant (3). According to [11], the assessor scores were averaged to produce a composite score and converted in a binary score by assuming as irrelevant $\langle newsaggregation, ad \rangle$ pairs with a composite score higher or equal to 2.34. We also calculated the variance, σ_k, and the average value, μ_k, of the assessor agreements, where k is the number of ads suggested by the proposed CA system for each news aggregation. Table 3 reports the results in terms of $p@k$, σ_k, and μ_k. Those results show that the $p@k$ is on average around 0.6. This is due to several issues. First, news aggregation descriptions are often too short and not enough informative for the assessor. Furthermore, let us note that some noise might be introduced by the fully automatic process of aggregation building. Moreover, some categories are very specific (e.g., *Religious Culture*) whereas others are very generic (e.g., *Economy and Finance*). To put into evidence this issue, the following columns in Table 3 (from **EF** to **T**) show the performances of the system in terms of $p@k$, σ_k, and μ_k, for each category.

Results clearly show that the more the category is specific, the better the system performs. To better highlight this point, Figure 4 shows the variance (σ_1) and the average value (μ_1) of assessor agreements considering only one advertisement for each category, i.e. setting $k = 1$. The green point is the average value μ_1 of the scores given by the assessors; whereas the red line shows the variance σ_1 around μ_1. Let us stress the fact that, for the sake of readability, we chose to display σ_1

Table 3 Results of the proposed CA system according to the assessor evaluation. The first column shows the average performance of the system, i.e. without making distinction among categories. The remaining columns show the performance for each of the considered categories.

	G	EF	ENT	HHS	MS	PPMM	RC	S	T
p@1	0.632	0.300	0.600	0.800	0.400	0.667	**0.900**	0.600	0.800
σ_1	0.411	0.197	0.508	0.573	0.359	0.473	0.381	0.386	0.414
μ_1	2.136	2.523	2.319	2.038	2.408	2.128	1.608	2.185	1.882
p@2	0.639	0.450	0.550	0.800	0.450	0.500	**0.900**	0.700	0.750
σ_2	0.429	0.309	0.459	0.566	0.381	0.441	0.410	0.395	0.472
μ_2	2.123	2.420	2.340	1.965	2.342	2.269	1.638	1.984	2.037
p@3	0.594	0.367	0.567	0.733	0.433	0.444	**0.867**	0.633	0.700
σ_3	0.414	0.330	0.431	0.528	0.374	0.405	0.385	0.367	0.492
μ_3	2.154	2.470	2.331	2.026	2.308	2.339	1.654	2.046	2.080
p@4	0.573	0.375	0.550	0.725	0.425	0.389	**0.850**	0.625	0.625
σ_4	0.400	0.338	0.428	0.520	0.340	0.367	0.368	0.361	0.477
μ_4	2.179	2.452	2.322	2.044	2.335	2.400	1.665	2.077	2.158
p@5	0.542	0.380	0.540	**0.720**	0.420	0.378	0.680	0.580	0.620
σ_5	0.391	0.331	0.432	0.528	0.330	0.342	0.342	0.362	0.459
μ_5	2.224	2.450	2.300	2.051	2.362	2.458	1.882	2.141	2.170

around μ_1 just to give a visual hint. The Figure shows that the shorter the red line is, the greater the agreement is. For instance, for *Economy and Finance*, assessors agree that the suggested ads are mostly *irrelevant*. On the other hand, for *Health and Health Services*, even if the average value is 2.04 (i.e., *somewhat relevant*), assessors are in disagreement.

8 The Proposed Approach in the Reference Scenario

To better illustrate how the proposed solutions work in practice, let us go back to the reference scenario sketched in Section 3. Suppose that Edinson is searching for information regarding the football match "Real Madrid - Barcelona" and he finds the news aggregation whose description is shown in Figure 5[5].

According to our proposal solution, the system analyzes all the news that compose the aggregation (in this case 42) and then each of them is summarized according to the TF2P technique and classified by the centroid-based classifier. All the *BoW* and the *CF* are collected by the *BoW Aggregator* and the *CF Aggregator*, respectively. Thanks to this approach, to evaluate if reading all the aggregated news, Edinson could first consider the suggested keywords shown in Figure 6.

---

[5] Currently, we are considering only Italian news, for the sake of clarity we translated the news in English.

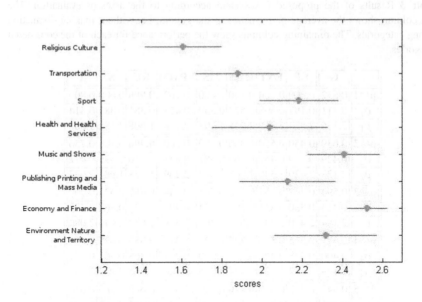

Fig. 4 The variance (σ_1) and the average value (μ_1) of assessor agreements for each category and for $k = 1$.

"Merengues" recovered after a sending off, but distance from the leadership of "Blaugrana" remains of 8 points. The goals of Messi and Cristiano Ronaldo, both by penalty kick, opened the row of "clasicos", in Liga, Copa del Rey, and Champions.
The final score of 1-1 delivers the Spanish championship to "Blaugrana", which holds a distance of 8 points from the rivals, but they failed to maintain the row of 5 successive wins of "clasicos", although the asset of one player and one goal. "Merengues" tied with a proud response, useless for winning the Liga, but that morally can be considered as a victory, since it happened after a row of defeats and bad performances.

Fig. 5 An example of description of a news aggregation.

champions
Barcelona
supporters
Ronaldo
extratime
Blaugrana
team
triumph
Liga
Madrid
Real
goal
Mourinho
catalan

Fig. 6 A selection of the keywords extracted from the news aggregation in Figure 5.

Then, if Edinson decides to read the news, the system calculates, for each ad in the ads repository, the matching with the given news aggregation according to the equation 1. In so doing, the system proposes to him the 5 ads depicted in Figure 7. As shown in the Figure, all the suggested ads are related to the main category *Sport*: a kit to synchronize your runner shoes with your mp3 player; two online sport newspapers; a website that sells sport and show tickets; and a video-game that allows users to make sports.

Fig. 7 The suggested ads.

9 Conclusions

In this chapter, we address two main issues: extracting relevant keywords to news and news aggregations, and automatically associating suitable advertisements to aggregated data. To achieve the first goal, we propose a solution based on the adoption of extraction-based text summarization techniques; whereas to achieve the second goal, we developed a contextual advertising system that works on multimodal aggregated data. To assess the proposed solutions, we performed experiments on Italian news aggregations. As for the extraction of keywords, results, calculated in terms of precision, recall, and F1, shown that the best performances are obtained when using the TF2P technique for both news and news aggregations. In other words, the best set of keywords is obtained considering the title, the first and second paragraph of each news. As for the association of ads, results, calculated in terms of $p@k$, clearly show that the more the category is specific, the better the system performs.

References

1. Anagnostopoulos, A., Broder, A.Z., Gabrilovich, E., Josifovski, V., Riedel, L.: Just-in-time contextual advertising. In: CIKM 2007: Proceedings of the Sixteenth ACM Conference on Conference on Information and Knowledge Management, pp. 331–340. ACM, New York (2007), doi:http://doi.acm.org/10.1145/1321440.1321488
2. Armano, G., Giuliani, A., Messina, A., Montagnuolo, M., Vargiu, E.: Experimenting text summarization on multimodal aggregation. In: Lai, C., Semeraro, G., Vargiu, E. (eds.) 5th International Workshop DART 2011, New Challenges on Information Retrieval and Filtering, CEUR Workshop Proceedings, vol. 771 (2011)
3. Armano, G., Giuliani, A., Vargiu, E.: Experimenting text summarization techniques for contextual advertising. In: IIR 2011: Proceedings of the 2nd Italian Information Retrieval (IIR) Workshop (2011)
4. Armano, G., Giuliani, A., Vargiu, E.: Semantic enrichment of contextual advertising by using concepts. In: International Conference on Knowledge Discovery and Information Retrieval (2011)
5. Armano, G., Giuliani, A., Vargiu, E.: Studying the impact of text summarization on contextual advertising. In: 8th International Workshop on Text-based Information Retrieval (2011)
6. Baxendale, P.: Machine-made index for technical literature - an experiment. IBM Journal of Research and Development 2, 354–361 (1958)
7. Brandow, R., Mitze, K., Rau, L.F.: Automatic condensation of electronic publications by sentence selection. Information Processing Management 31, 675–685 (1995)
8. Broder, A., Fontoura, M., Josifovski, V., Riedel, L.: A semantic approach to contextual advertising. In: SIGIR 2007: Proceedings of the 30th Annual International ACM SIGIR Conference on Research and Development in Information Retrieval, pp. 559–566. ACM, New York (2007), doi:http://doi.acm.org/10.1145/1277741.1277837
9. Broder, A.Z., Ciaramita, M., Fontoura, M., Gabrilovich, E., Josifovski, V., Metzler, D., Murdock, V., Plachouras, V.: To swing or not to swing: learning when (not) to advertise. In: Shanahan, J.G., Amer-Yahia, S., Manolescu, I., Zhang, Y., Evans, D.A., Kolcz, A., Choi, K.S., Chowdhury, A. (eds.) CIKM, pp. 1003–1012. ACM (2008)

10. Ciaramita, M., Murdock, V., Plachouras, V.: Online learning from click data for sponsored search. In: Proceeding of the 17th International Conference on World Wide Web, WWW 2008, pp. 227–236. ACM, New York (2008)
11. Ciaramita, M., Murdock, V., Plachouras, V.: Semantic associations for contextual advertising. Journal of Electronic Commerce Research 9(1), 1–15 (2008); Special Issue on Online Advertising and Sponsored Search
12. Das, D., Martins, A.F.: A survey on automatic text summarization. Tech. Rep. Literature Survey for the Language and Statistics II course at CMU (2007)
13. Deschacht, K., Moens, M.-F.: Finding the Best Picture: Cross-Media Retrieval of Content. In: Macdonald, C., Ounis, I., Plachouras, V., Ruthven, I., White, R.W. (eds.) ECIR 2008. LNCS, vol. 4956, pp. 539–546. Springer, Heidelberg (2008)
14. Edmundson, H.P.: New methods in automatic extracting. Journal of ACM 16, 264–285 (1969)
15. Kokar, M.: Formalizing classes of information fusion systems. Information Fusion 5(3), 189–202 (2004)
16. Kołcz, A., Alspector, J.: Asymmetric missing-data problems: Overcoming the lack of negative data in preference ranking. Information Retrieval 5, 5–40 (2002)
17. Kolcz, A., Prabakarmurthi, V., Kalita, J.: Summarization as feature selection for text categorization. In: CIKM 2001: Proceedings of the Tenth International Conference on Information and Knowledge Management, pp. 365–370. ACM, New York (2001)
18. Lacerda, A., Cristo, M., Gonçalves, M.A., Fan, W., Ziviani, N., Ribeiro-Neto, B.: Learning to advertise. In: SIGIR 2006: Proceedings of the 29th Annual International ACM SIGIR Conference on Research and Development in Information Retrieval, pp. 549–556. ACM, New York (2006), doi:doi.acm.org/10.1145/1148170.1148265
19. Laudy, C., Ganascia, J.-G.: Information fusion in a tv program recommendation system. In: 11th International Conference on Information Fusion 2008, pp. 1–8 (2008)
20. Li, X., Yan, J., Deng, Z., Ji, L., Fan, W., Zhang, B., Chen, Z.: A novel clustering-based RSS aggregator. In: Proc. of WWW 2007, pp. 1309–1310 (2007)
21. Liu, H., Singh, P.: Conceptnet: A practical commonsense reasoning tool-kit. BT Technology Journal 22, 211–226 (2004)
22. Luhn, H.P.: The automatic creation of literature abstracts. IBM Journal of Research and Development 2, 159–165 (1958)
23. Mahler, R.P.S.: Statistical Multisource-Multitarget Information Fusion. Artech House, Inc., Norwood (2007)
24. Mani, I.: Automatic summarization. John Benjamins, Amsterdam (2001)
25. Messina, A., Borgotallo, R., Dimino, G., Gnota, D.A., Boch, L.: Ants: A complete system for automatic news programme annotation based on multimodal analysis. In: Intl. Workshop on Image Analysis for Multimedia Interactive Services (2008)
26. Messina, A., Montagnuolo, M.: Multimodal aggregation and recommendation technologies applied to informative content distribution and retrieval. In: Soro, A., Vargiu, E., Armano, G., Paddeu, G. (eds.) Information Retrieval and Mining in Distributed Environments. SCI, vol. 324, pp. 213–232. Springer, Heidelberg (2010)
27. Messina, A., Montagnuolo, M.: Heterogeneous data co-clustering by pseudo-semantic affinity functions. In: Proc. of the 2nd Italian Information Retrieval Workshop, IIR (2011)
28. Murdock, V., Ciaramita, M., Plachouras, V.: A noisy-channel approach to contextual advertising. In: Proceedings of the 1st International Workshop on Data Mining and Audience Intelligence for Advertising, ADKDD 2007, pp. 21–27. ACM, New York (2007)
29. Nenkova, A.: Automatic text summarization of newswire: lessons learned from the document understanding conference. In: Proceedings of the 20th National Conference on Artificial Intelligence, vol. 3, pp. 1436–1441. AAAI Press (2005)

30. Nguyen, L.-D., Woon, K.-Y., Tan, A.-H.: A self-organizing neural model for multimedia information fusion. In: 11th International Conference on Information Fusion, pp. 1–7 (2008)

31. Radev, D.R.: A common theory of information fusion from multiple text sources step one: cross-document structure. In: Proceedings of the 1st SIGdial Workshop on Discourse and Dialogue, pp. 74–83. Association for Computational Linguistics, Morristown (2000)

32. Radev, D.R., Hovy, E., McKeown, K.: Introduction to the special issue on summarization. Computational Linguistic 28, 399–408 (2002)

33. Ribeiro-Neto, B., Cristo, M., Golgher, P.B., Silva de Moura, E.: Impedance coupling in content-targeted advertising. In: SIGIR 2005: Proceedings of the 28th Annual International ACM SIGIR Conference on Research and Development in Information Retrieval, pp. 496–503. ACM, New York (2005),
doi:http://doi.acm.org/10.1145/1076034.1076119

34. Rocchio, J.: Relevance feedback in information retrieval. In: The SMART Retrieval System: Experiments in Automatic Document Processing, pp. 313–323. Prentice Hall (1971)

35. Salton, G., McGill, M.: Introduction to Modern Information Retrieval. McGraw-Hill Book Company (1984)

36. Wu, Y., Chang, E.Y., Chang, K.C.-C., Smith, J.R.: Optimal multimodal fusion for multimedia data analysis. In: MULTIMEDIA 2004: Proceedings of the 12th Annual ACM International Conference on Multimedia, pp. 572–579. ACM, New York (2004)

37. Xu, C., Wang, J., Lu, H., Zhang, Y.: A novel framework for semantic annotation and personalized retrieval of sports video. IEEE Trans. on Multimedia 10(3), 421–436 (2008)

38. Yih, W.t., Goodman, J., Carvalho, V.R.: Finding advertising keywords on web pages. In: WWW 2006: Proceedings of the 15th International Conference on World Wide Web, pp. 213–222. ACM, New York (2006),
doi: http://doi.acm.org/10.1145/1135777.1135813

ImageHunter: A Novel Tool for Relevance Feedback in Content Based Image Retrieval

Roberto Tronci, Gabriele Murgia, Maurizio Pili, Luca Piras, and Giorgio Giacinto

Abstract. Nowadays, a very large number of digital image archives is easily produced thanks to the wide diffusion of personal digital cameras and mobile devices with embedded cameras. Thus, personal computers, personal storage units, as well as photo-sharing and social-network websites, are rapidly becoming the repository for thousands, or even billions of images (i.e., more than 100 million photos are uploaded every day on the social site Facebook). As a consequence, there is an increasing need for tools enabling the semantic search, classification, and retrieval of images. The use of meta-data associated to images solves the problems only partially, as the process of assigning reliable meta-data to images is not trivial, is slow, and closely related to whom performed the task. One solution for effective image search and retrieval is to combine content-based analysis with feedbacks from the users. In this chapter we present Image Hunter, a tool that implements a Content Based Image Retrieval (CBIR) engine with a Relevance Feedback mechanism. Thanks to a user friendly interface the tool is especially suited to unskilled users. In addition, the modular structure permits the use of the same core both in web-based and stand alone applications.

1 Introduction

The growing number of digital data such as text, video, audio, pictures or photos is pushing the need for tools allowing the quick and accurate retrieval of information

Roberto Tronci · Gabriele Murgia · Maurizio Pili
AmILAB - Laboratorio Intelligenza d'Ambiente, Sardegna Ricerche, loc. Piscinamanna,
I-09010 Pula(CA), Italy
e-mail: {roberto.tronci,gabriele.murgia}@sardegnaricerche.it,
        maurizio.pili@sardegnaricerche.it

Luca Piras · Giorgio Giacinto
DIEE - Department of Electric and Electronic Engineering, University of Cagliari,
Piazza d'Armi snc, I-09123 Cagliari, Italy
e-mail: {luca.piras,giacinto}@diee.unica.it

C. Lai et al. (Eds.): New Challenges in Distributed Inf. Filtering and Retrieval, SCI 439, pp. 53–70.
springerlink.com   © Springer-Verlag Berlin Heidelberg 2013

from data. Whereas the results of traditional text data search methods are quite satisfactory, the same can not be said for visual or multimedia data. So far, the most common method for image retrieval is predicated on adding meta-data to images as keywords, tag, label or short descriptions, so that the retrieval can occur through such annotations. The manual cataloging of images, even though it requires expensive work and a large amount of time, is often not so effective. Describing a picture in words is not always easy, and the relevance of the description is strictly subjective.

By now, all mobile phones are equipped with cameras, and thanks to the widespread use of the Web, social networks and almost "unlimited" storage space, the exchange of photos and digital images has become frenetic, to say the least. As a consequence there is an increasing need for tools enabling the semantic search, classification, and retrieval of images. As above-mentioned, the use of meta-data associated to the images solves the problems only partly, as the process of assigning meta-data to images is not trivial, slow, and closely related to the persons who performed the task. This is especially true for retrieval tasks in very highly populated archives, where images exhibit high variability in semantic. It turns out that the description of image content tends to be intrinsically subjective and partial, and the search for images based on keywords may fit users' needs only partially. For this reason, since the early nineties, the scientific community focused on the study of Content Based Image Retrieval [13, 19, 15, 8] that it is based on the idea of indexing image by using low-level features such as color, texture, shape, etc.. Another difficulty in devising effective image retrieval and classification tools is given by the vast amount of information conveyed by images, and the related subjectivity of the criteria to be used to assess the image content. This kind of problem is called *semantic gap* [19] and it is precisely due to the different ways in which human beings and machines interpret the images. For the humans, these arouse emotions, memories or also reflections; for a computer are simple sets of pixels from which to extract numerical values. In order to capture such subjectivity, image retrieval tools may employ the so called *relevance feedback* [17, 25]. Relevance feedback techniques involve the user in the process of refining the search. In a CBIR task in which the RF is applied, the user submits a query image to the system, that is an example of the pictures of interest; starting from the query, the system assigns a score to the images in the database, the score being related to a similarity measure between the images and the query. A number of best scored images are returned to the user that judges them as relevant or not. This new information is exploited by the system to improve the search and provide a more accurate result in the next iteration. Faced with this new scenario, it has become increasingly urgent to find a way to manage this heap of data, to permit an effective search and to involve the user in this task.

Image Hunter is a full content-based image retrieval tool which does not need a text query in contrast to the vast majority of other applications [18, 2]. It is able to retrieve an ensemble of "similar" images from an image archive starting from an image provided by a user. Image Hunter is further equipped with a learning mechanism based on the relevance feedback paradigm that allows dynamically adapting and refining the search. In addition, the adaptability of the system has been enforced by the concurrent use of twelve different feature sets including colour based, texture, and shape global descriptors.

The rest of the chapter is organized as follows. Section 2 illustrates some related image retrieval systems. Section 3 shows how Image Hunter is organized and explains how it works, in particular Section 3.3 briefly reviews the integrated learning process and relevance feedback mechanisms implemented in the application. Experimental results are reported in Section 4. Conclusions are drawn in Section 5.

2 Related Works

A rigorous comparison of image retrieval engines is not easy to be performed out of specific competitions where all participants have to use the same dataset, the same query modality, and possibly the same feature sets. For these reasons, this section proposes just a short review of a subset of this kind of systems. First of all, we summarize the characteristics that in our opinion can be used to categorize different image retrieval engines. The most important one is the approach used for image search that can be based either on *content*, or on *meta-data*, or on both. In the first case, only visual characteristics such as shapes,colours etc. are used in the retrieval process, whereas in the second case other information such as the name of the file, tags or labels given by users, GPS position, etc., are exploited. There are also systems that can use both information, even if they are usually characterized by a heavier computational cost, while providing better results w.r.t. systems based on single retrieval modality.

In order to improve the performances, and better satisfy the user's needs, the system can interact with her through *Relevance Feedback* mechanisms, i.e., by asking the user to judge one or more images as relevant or not to the query. Also in this case, the choice between one or more feedback sources implies a lower or higher complexity in the retrieval algorithm. Another feature that can be used to categorize an IR system, is the query modality. The query can be formulated either by text, or by example, and in the latter case, it is possible to further distinguish between examples submitted by the users, or examples chosen in a set of images.

As above mentioned, **Image Hunter** is a full content-based image retrieval tool equipped with a learning mechanism that is designed to provide multiple images Relevance Feedback functionalities. The query can be submitted by example, and the user can choose the query image between those memorized in the image archive of the system, and those stored in his own hard disk. In the following, we will provide a summary description of the most relevant image retrieval engines that exhibit functionalities similar to those offered by Image Hunter.

Anaktisi[1] is a web-site designed with particular attention to the size and storage requirements of the features image descriptors. The retrieval approach is of the content-based type, and adopts a single relevance feedback mechanism. The query can be chosen from the images proposed by the system or it can be uploaded by an user.

---

[1] http://orpheus.ee.duth.gr/anaktisi/

Alipr[2] is an image search engine that ask the user to upload an image, or to type the URL of an image on the web, and then returns to the user the most suitable tags for the image by selecting them from a pre-computed set. After this first step, the system shows some pictures that "may" be related to the query according to the selected tags. In the case the user is not satisfied with the results, *Alipr* suggests to search for visually similar pictures through the **SIMPLIcity**[3] search engine [24]. This CBIR system does neither exploit feedback information form the user, nor meta-data associated to images.

Pixolu[4] is an image search system that combines keyword search with content based search, and presents to the user inter-image relationships. After a first step in which the user must enter a term in a search box, the system searches the image indexes of Yahoo and Flickr. The returned images are then visually sorted, and the user can choose the images that are relevant to her desired search result. In the following step, the selected images are used to refine the results by filtering out visually non-similar images.

Lucignolo[5] is an image similarity search engine that can use both the native full-text search engine Lucene, and colour and texture features. It is able to accept both text and visual queries but does not involve the user in the refinement of the search.

Google Image is probably the best known image search engine. The service **Similar Images**, offered as an experimental tool since the end of 2009, has been "fine tuned" by *Google Labs*, and has become a permanent feature in Google Images. Differently to *Lucignolo*, Google Similar Images does not allow choosing between meta-data, content-based data, or both, but the user can just decide if the target of the search is to find similar images, or to find different sizes of the same input picture.

Table 1 summarizes the main characteristics of the image search engines reviewed.

3 Image Hunter

With the aim of building a practical application to show the potentialities of Content Based Image Retrieval tools with Relevance Feedback, we developed Image Hunter[6]. This tool is entirely written in JAVA, so that the tool is machine independent. For its development, we partially took inspiration from the LIRE library [16] (that is just a feature extraction library). In addition, we chose Apache Lucene[7] for building the index of the extracted data.

Image Hunter is made up of two main parts: the core, and the user interface (see Figure 1).

---

[2] http://alipr.com/

[3] http://alipr.com/cgi-bin/zwang/regionsearch_show.cgi

[4] http://www.pixolu.de/

[5] http://lucignolo.isti.cnr.it/

[6] http://prag.diee.unica.it/amilab/WIH

[7] http://lucene.apache.org/

Table 1 Image Retrieval search engines.

| Systems | Characteristics | | | | | |
	Content Based	Meta Data	Relevance Feedback	Multiple Relevance Feedback	Text Query	Private Images
Image Hunter	✓	✗	✓	✓	✗	✓
Anaktisi	✓	✗	✓	✗	✗	✓
ALIPR	✗	✓	✗	✗	✗	✓
SIMPLIcity	✓	✗	✗	✗	✗	✓
Pixolu	✓	✗	✓	✓	✓	✗
Lucignolo	✓	✓	✗	✗	✓	✓
Google Similar Images	✓	✓	✗	✗	✓	✓

Fig. 1 Image Hunter structure

The main core of Image Hunter is a full independent module, thus allowing the development of a personalized user interface. The core is subdivided into four parts:

- Indexing;
- Feature extraction interface;
- Lucene interface for data storing;
- Image Retrieval and Relevance Feedback.

We'll briefly describe how the indexing is made and it is stored by means of the Apache Lucene. After, we will describe the visual features that we embedded into

Image Hunter. Finally, we will describe the *Image Retrieval and Relevance Feedback* module that is the more important module as it implements the core engine of our system. Each time a user submits an image to be used as a *visual query*, the system computes the visual similarity between the query and each image in the collection. The visual similarity is computed in terms of the average value of the normalized distances in each feature space. Then, the user can label the images provided by the system as relevant to her search or not, and the system exploits this feedback to learn which is the best combination of visual features that represents the semantic meaning that the user is associating to the query. Thus, in the feedback elaboration process, the visual similarity is computed in terms of a weighted combination of the distances in different feature spaces, rather than in terms of the average distance. In the following sections we describe the Relevance Feedback techniques implemented in Image Hunter (Section 3.3), and the web-based user interface that we have developed (Section 3.4).

3.1 Indexing and Data Storing

The *Indexing* part has the role of extracting the visual features and other informations from the images. The visual features and other descriptors of the images are then stored in a particular structure defined inside Image Hunter. The tool can index different types of image formats and it can be built in an incremental way. For each image collection, all the data are stored in a database built according the Apache Lucene standard. "Apache Lucene is a high-performance, full-featured text search engine library written entirely in Java. It is a technology suitable for nearly any application that requires full-text search, especially cross-platform." The core of Lucene's logical architecture is the idea of a document containing text fields. Moreover Lucene is an Open-Source project.

Lucene turned out to be well suited for the storage needs of Image Hunter, as it resulted faster than SQL based solutions that we have tested. The core of Lucene's logical architecture is a series of document containing text fields, where we have associated different features (fields) to each image (document). In particular, the indexes created according to the Lucene model can be easily moved by copying the folder that contains the index. In this way it is also quite simple to build a "portable" version of Image Hunter. Moreover, we had also adapted some of the main classes defined by Lucene to better fit our needs: e.g., we created some methods to simplify the index administration, and we enriched the functionality of document manipulations.

3.2 Feature Extraction Interface

As we said before, for the feature extraction we took inspiration from the LIRE library that it is used as an external feature extraction library. We expanded and modified its functionalities by implementing or reimplementing in Image Hunter some extractors.

The *Feature extraction interface* allows to extract different visual features based on different characteristics: colour, texture and shape. They are:

- *Scalable Color* [4], a colour histogram extracted from the HSV colour space;
- *Color Layout* [4], that characterizes the spatial distribution of colours;
- *RGB-Histogram* and *HSV-Histogram* [16], based on RGB and HSV components of the image respectively;
- *Fuzzy Color* [16], that considers the colour similarity between the pixel of the image;
- *JPEG Histogram* [16], a JPEG coefficient histogram;
- *Appearance-Based Image Features* [9] obtained rescaling the images to 32x32 size and returning a colour histogram extracted from the RGB colour space;
- *Edge Histogram* [4], that captures the spatial distribution of edges;
- *Tamura* [20], that captures different characteristic of the images like coarseness, contrast, directionality, regularity, roughness;
- *Gabor*[9] that allows the edge detection;
- *CEDD* (Color and Edge Directivity Descriptor) [5];
- *FCTH* (Fuzzy Color and Texture Histogram) [6].

3.3 Relevance Feedback Techniques Implemented in Image Hunter

In this section the three relevance feedback techniques implemented in the core are described. Two of them are based on the nearest-neighbour paradigm, while one of them is based on Support Vector Machines. The use of the nearest-neighbour paradigm is motivated by its use in a number of different pattern recognition fields, where it is difficult to produce a high-level generalization of a class of objects, but where neighbourhood information is available [1, 10]. In particular, nearest-neighbour approaches have proven to be effective in outliers detection, and one-class classification tasks [3, 21]. Support Vector Machines are used because they are one of the most popular learning algorithm when dealing with high dimensional spaces as in CBIR [7, 22].

3.3.1 k-NN Relevance Feedback

In this work we resort to a technique proposed by some of the authors in [11] where a score is assigned to each image of a database according to its distance from the nearest image belonging to the target class, and the distance from the nearest image belonging to a different class. This score is further combined to a score related to the distance of the image from the region of relevant images. The combined score is computed as follows:

$$rel(\mathbf{I}) = \left(\frac{n/t}{1+n/t}\right) \cdot rel_{BQS}(\mathbf{I}) + \left(\frac{1}{1+n/t}\right) \cdot rel_{NN}(\mathbf{I}) \qquad (1)$$

where n and t are the number of non-relevant images and the whole number of images retrieved after the latter iteration, respectively. The two terms rel_{NN} and rel_{BQS} are computed as follows:

$$rel_{NN}(\mathbf{I}) = \frac{\|\mathbf{I} - NN^{nr}(\mathbf{I})\|}{\|\mathbf{I} - NN^{r}(\mathbf{I})\| + \|\mathbf{I} - NN^{nr}(\mathbf{I})\|} \tag{2}$$

where $NN^{r}(\mathbf{I})$ and $NN^{nr}(\mathbf{I})$ denote the relevant and the non relevant Nearest Neighbour of $\mathbf{I}$, respectively, and $\|\cdot\|$ is the metric defined in the feature space at hand,

$$rel_{BQS}(\mathbf{I}) = \frac{1 - e^{1 - d_{BQS}(\mathbf{I})\big/\max_i d_{BQS}(\mathbf{I}_i)}}{1 - e} \tag{3}$$

where e is the *Euler's number*, i is the index of all images in the database and d_{BQS} is the distance of image $\mathbf{I}$ from a reference vector computed according to the Bayes decision theory (Bayes Query Shifting, BQS) [12]. If we are using F feature spaces, we have different scores $rel(\mathbf{I})$ for each f feature space. Thus the following combination is performed to obtain a "single" score:

$$rel(\mathbf{I}) = \sum_{f=1}^{F} w_f \cdot rel^f(\mathbf{I}) \tag{4}$$

where the w_f is the weight associated to the f-space. In this chapter we are going to use two ways of computing the weights w_f. One approach to estimate the weights w_f is to take into account the minimum distance between all the pairs of relevant images, and the minimum distance between all the pairs of relevant and non-relevant images as follows

$$w_f = \frac{\sum_{i \in R} d_{min}^f(\mathbf{I}_i, R)}{\sum_{i \in R} d_{min}^f(\mathbf{I}_i, R) + \sum_{i \in R} d_{min}^f(\mathbf{I}_i, N)} \tag{5}$$

The other approach for estimating the weights w_f, is a modification of the previous one. Let us sort the images according to their distances from the query as measured by $rel(\mathbf{I})$, then their rank, from the closer to the farther, is considered. The weights are then computed by taking into account the relevant images and their "positions" in a f-space, and the sum of all the "positions" in all the feature spaces F as follows

$$w_f = \frac{\sum_{i=1}^{R} \frac{1}{pos_i^f}}{\sum_{k=1}^{F} \sum_{i=1}^{R} \frac{1}{pos_i^k}} \tag{6}$$

3.3.2 SVM Based Relevance Feedback

Support Vector Machines are used to find a decision boundary in each feature space $f \in F$. The SVM is very handy for this kind of task because, in the case of image retrieval, we deal with high dimensional feature spaces and two "classes" (i.e. relevant and not-relevant). For each feature space f, a SVM is trained using the feedback given by the user. The results of the SVMs in terms of distances from the hyperplane of separation are then combined into to a relevance score through the Mean rule as follows

$$rel_{SVM}(\mathbf{I}) = \frac{1}{F} \sum_{f=1}^{F} rel_{SVM}^{f}(\mathbf{I}) \qquad (7)$$

3.4 Image Hunter's User Interface

The user interface is structured to provide just the functionalities that are strictly related with the user interaction (e.g., the list of relevant images found by the user). Image Hunter employs a web-based interface that can be viewed at the address *http://prag.diee.unica.it/amilab/WIH*. This version is a web application built for the Apache Tomcat web container by using a mixture of JSP and java Servlet. The graphic interface is based on the jQuery framework, and has been tested for the Mozilla Firefox and Google Chrome browsers. When the web container is launched, a servlet checks if the collection contains new images to keep the Lucene index up-to-date. Afterwards, the index is loaded by the web application and used for all the sessions opened by the remote clients (see Figure 2). The Image Hunter homepage let the user choose the picture from which starting the search. The picture can be chosen either within those of the proposed galleries or among the images from the user hard disk (see Figure 3). Each query is managed by a servlet that queries the Image Hunter engine and displays the 23 most similar images according

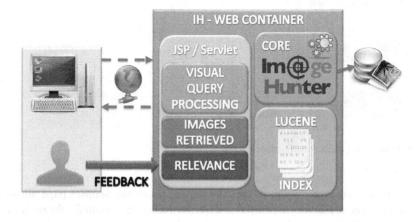

Fig. 2 Web Application Architecture

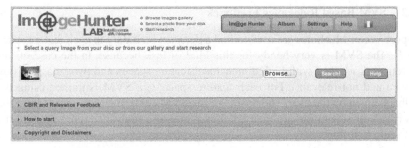

Fig. 3 Web Application home page

to the mechanisms reported in Section 3.3. The choice of the number of images
displayed to the user on the one hand takes into account the needs of the page layout
and, on the other hand, is oriented to keep a high level of users' attention. In order
to make intuitive and easy the features offered by the application, the graphical
interface has been designed relying on the *Drag and Drop* approach (see Figure
4). From the result page, the user can drag the images that her deems relevant to
her search in a special boxcart, and then submit the feedback. Then the feedback is
processed by the system, and a new set of images is proposed to the user. The user
can then perform another feedback round.

Fig. 4 Results and Relevance Feedback

In order to make the system flexible for skilled users, the *Settings* page allows
choosing the visual features to be used to describe the image content for the retrieval
process (by default all the visual features are used).

One of Image Hunter's greatest strengths is its flexibility: in fact, its structure was built in a way that it is possible to add any other image descriptor. The choice of the above mentioned set is due to the "real time" nature of the system with large database. In fact even if some local features such as SIFT or SURF could improve the retrieval performances for some particular kind of searches, on the other hand they are more time expensive in the evaluation of the similarity between images.

In addition, in the *Settings* page the user can select the Relevance Feedback technique to be used, and the dataset to explore.

Finally, Figure 5 summarizes the typical user interaction within *Image Hunter*.

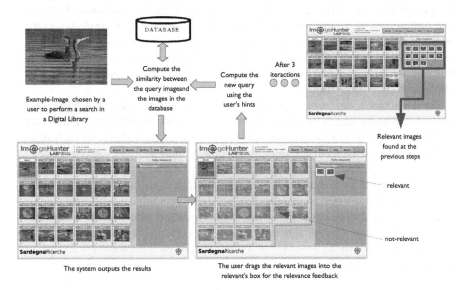

Fig. 5 Example of a typical user interaction with *ImageHunter*

4 Experiments

4.1 Dataset Setup

In the experimental evaluation of Image Hunter we performed both a full automatic test by using the MIRFlickr-25000 collection [14] and a *User Experience* test using a set of 53279 unlabelled images extracted from the *Sardegna Digital Library*[8] through the time. MIRFlickr-25000 consists of 25000 images tagged by the user of the social photography site Flickr. The average number of tags per image is 8.94. In the collection there are 1386 tags which occur in at least 20 images. Moreover, for a limited number of images, some manual annotations is also available (24 annotations in the collection considered for this experiment). In these experiments we used all the features embedded in the system that have been listed in Section 3.2.

---

[8] http://www.sardegnadigitallibrary.it

In the automatic test, we analysed all the tags of the collection by a semantic point of view, and fused the tags with the annotations in a tag verification process (i.e., we have verified if there were different tag names used for the same concept). This process was performed to keep only the tags which occur in at least 100 images, so that the single tags/concepts are adequately represented in the dataset used in the evaluation experiments. This process of fusing and discarding tags brought us to keep 24718 images and 69 tags, with an average number of tags per image of 4.19. Thus, as query images, we chose 1294 of them from the refined collection. These query images have a number of tags per image that varies from 3 to 10 (i.e., the single image can represent different meanings), with an average number of tags per image equal to 4.69 (thus very similar to the value in all the collection). For each one of the 1294 query images, we considered one at a time each single tag as the target of the retrieval process to be refined through the relevance feedback. Thus, each query image has been used as starting example for different retrieval tasks. In this way, 6070 retrieval tasks were performed for each relevance feedback technique implemented in *Image Hunter*.

Each automatic experiment consists of 10 iterations: the first one is based on a nearest neighbour search on all the feature spaces, and the other 9 iterations are based on one of the relevance feedback techniques described above. At each iteration we simulated the feedback from the user on 20 images.

The *User Experience* test has been performed by about 50 users that were asked to perform one or more searches by choosing as query one out of 32 images (See Figure 6) that we selected so that they exhibited different subjects, different colours and shapes. The users can choose to perform any number of consecutive iterations to refine the search. On average, each of the 32 queries has been used 6.75 times and the users performed an average of 5 iterations. At each iteration $n = 23$ images are shown to the user for marking the feedback.

Fig. 6 User Experience queries

4.2 Performance Measurements

The performance of the experiments will be assessed using the *Precision* and a modified definition of the Recall, that we named "user perceived" *Recall*.

The *Precision* is a measure that captures how many relevant images are found within the images that are "shown" to a user, and it is computed as follows:

$$p = \frac{A(q) \cap R(q)}{A(q)} \tag{8}$$

where $A(q)$ is the ensemble of images retrieved by using the query q, while $R(q)$ is the ensemble of images that in the collection are relevant according to the query q.

The *Recall* measures how many relevant images are found among the set of images in the collection that have the same tag/concept:

$$r = \frac{A(q) \cap R(q)}{R(q)} \tag{9}$$

In this way we compute the percentage of relevant images with respect to the total number of relevant images in the collection. This measure has a disadvantage: if the total number of relevant images in a collection for a given tag is greater than the number of images shown by the system, the measure is going to be always less than 100% even if all the images shown to the user are relevant. Thus, this measure doesn't represent the perception, in term of performance, that a real user will have on the system. In addition, each class contains a different number of images, and therefore the denominator of Equation 9 differs from one class to another even in one order of magnitude, and it can completely distort the average performance. For these reasons, we propose to use a modification of the recall measure namely, the "user perceived" *Recall*. This measure takes into account just the maximum number of relevant images that can be shown to the user according to the number of iterations, and the number of images displayed per iteration, and it is computed as follows

$$r_p = \frac{A(q) \cap R(q)}{R^*(q)} \quad , \quad R^*(q) = \begin{cases} R(q) & \text{, if } |R(q)| \leq n \cdot i \\ n \cdot i & \text{, otherwise} \end{cases}$$

where $A(q)$ is the number of images retrieved by using the query q up to the iteration i, $R(q)$ is the number of relevant images in the dataset (for the query target), $|\cdot|$ indicates the cardinality of the set, and n is the number of images shown to the user per iteration.

In an unlabelled dataset it is more difficult to define the set of similar images, so in the *User Experience* test for each query image we have formed the set of relevant images by considering the images marked as Relevant by more than the 50% of users. Indicating as $\widehat{R}(q)$ this ensemble of images that are relevant to the query q the *Recall* will be:

$$r = \frac{A(q) \cap \widehat{R}(q)}{\widehat{R}(q)}. \tag{10}$$

It is worth to note that $|\widehat{R}(q)| \leq n \cdot i$, so the *Recall* and the "user perceived" *Recall* agree.

4.3   *Experimental Results*

In the automatic test we compared the performance of all the relevance feedback techniques described in the previous section, i.e., the k-NN based on Equation (5) (*NN* in the tables), the k-NN based on Equation (6) (*PR* in the table), and the SVM, with the performance attained by simply *browsing* the image collection.

The term *browsing* indicates nothing more than showing the user the n images nearest to the query with no feedback [23]. The aim of comparing relevance feedback with *browsing* is to show the benefits of relevance feedback. To put it simple: can a relevance feedback approach retrieve more relevant images than simply *browsing* the collection by sorting the images according to the visual similarity with the query?

The average results in terms of *Precision*, and "user perceived" *Recall* obtained in the automatic test are presented in Figures 7 and 8, respectively. The results show that, as the number of iterations increase, the performance of the relevance feedback methods increase, as well as the difference in performance with *browsing*. From these analysis it turns out that the behaviour of the two k-NN methods are quite similar, while the SVM exhibits the biggest increasing performance power.

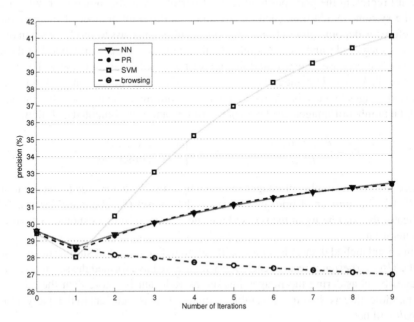

Fig. 7 *Precision* in the MIRFlickr experiments.

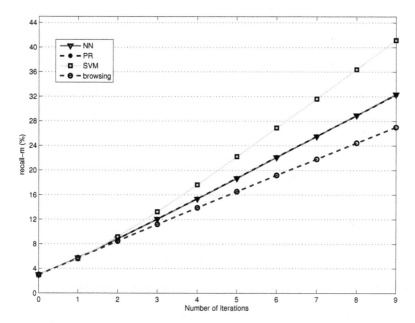

Fig. 8 "User perceived" *Recall* in the MIRFlickr experiments.

In the *User Experience* test, we used only the NN and the SVM relevance feed-back mechanism. In this scenario we marked as "Relevant" all those images considered Relevant by more than the 50% of the user in a certain iteration, given a relevance feedback mechanism. In this case, we show only the values of the recall measure proposed above as it allows to capture the performance of the system as it is perceived by real users. Figure 9 reports the obtained results. The performance of the relevance feedback techniques with respect to *browsing* shows that user interaction permits a very big improvement of the performance, and that the system learns how to find images that fulfil the user's desires.

We observed that users tend to label as "Relevant" less and less images after few iterations, especially if he/she is satisfied with the previous results, because labelling the images is an annoying task. As a consequence, the reported values of the recall can be considered as a lower bound of the true performances, as they take into account just the images actually labelled by the user. This aspect is captured by Figure 10 where the percentage average number of users per image at each interaction is showed: e.g., at interaction 5 only the 51, 11% and 43, 24% of initial users is still using the system in the case of NN and SVM respectively. The average number rapidly decrease with the increase of interactions, that's why in Figure 9 we showed only the performances between 1 and 5 (from 6 they are statistically meaningless).

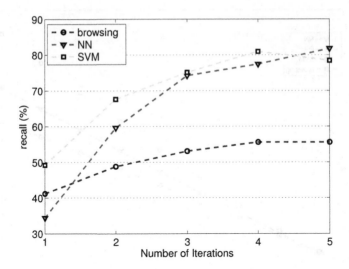

Fig. 9 *Recall* in the SDL experiments for the Relevance Feedback interactions.

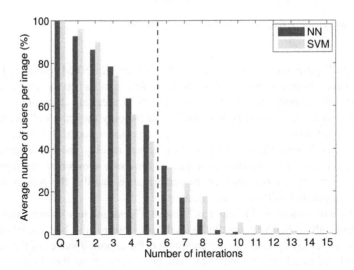

Fig. 10 Average number of users per image at each interaction.

5   Conclusions and Future Work

In this chapter we presented Image Hunter, a tool that exploits the potentiality of
Relevance Feedback to improve the performance of Content Based Image Retrieval.
Unlike other proposed tools, Image Hunter is a full content based image retrieval
system in which the user's feedback is integrated in the core of the application,

and permits a dynamical adaptation of the queries driven by the user. The proposed results obtained both in a full automatic test, and in a user test show how the integration of the relevance feedback improves significantly the performance of the image retrieval system making the search more effective with respect to *browsing*.

In the near future we are planing to extend our tool by adding content based classifiers (e.g., indoor/outdoor classifiers, face detectors, etc...) and study how they can be exploited in the actual relevance feedback system or by developing a totally new relevance feedback system able to exploit the user's feedback for similarity retrieval and classification.

References

1. Aha, D.W., Kibler, D., Albert, M.K.: Instance-based learning algorithms. Machine Learning 6(1), 37–66 (1991)
2. Barthel, K.U.: Improved image retrieval using automatic image sorting and semi-automatic generation of image semantics. In: WIAMIS 2008: Proceedings of the 2008 Ninth International Workshop on Image Analysis for Multimedia Interactive Services, pp. 227–230. IEEE Computer Society Press, Washington, DC (2008),
   doi:http://dx.doi.org/10.1109/WIAMIS.2008.56
3. Breunig, M.M., Kriegel, H.P., Ng, R.T., Sander, J.: LOF: Identifying density-based local outliers. In: W. Chen, J.F. Naughton, P.A. Bernstein (eds.) SIGMOD Conference, pp. 93–104. ACM (2000),
   doi: http://doi.acm.org/10.1145/342009.335388,db/
   conf/sigmod/BreunigKNS00.html
4. Chang, S.F., Sikora, T., Puri, A.: Overview of the mpeg-7 standard. IEEE Trans. Circuits Syst. Video Techn.
5. Chatzichristofis, S.A., Boutalis, Y.S.: CEDD: Color and Edge Directivity Descriptor: A Compact Descriptor for Image Indexing and Retrieval. In: Gasteratos, A., Vincze, M., Tsotsos, J.K. (eds.) ICVS 2008. LNCS, vol. 5008, pp. 312–322. Springer, Heidelberg (2008)
6. Chatzichristofis, S.A., Boutalis, Y.S.: Fcth: Fuzzy color and texture histogram - a low level feature for accurate image retrieval. In: Proceedings of the 2008 Ninth International Workshop on Image Analysis for Multimedia Interactive Services, pp. 191–196. IEEE Computer Society (2008), doi:10.1109/WIAMIS.2008.24
7. Cristianini, N., Shawe-Taylor, J.: An Introduction to Support Vector Machines and Other Kernel-based Learning Methods. Cambridge University Press (2000)
8. Datta, R., Joshi, D., Li, J., Wang, J.Z.: Image retrieval: Ideas, influences, and trends of the new age. ACM Computing Surveys 40(2), 1–60 (2008),
   doi:http://doi.acm.org/10.1145/1348246.1348248
9. Deselaers, T., Keysers, D., Ney, H.: Features for image retrieval: an experimental comparison. Inf. Retr. 11(2), 77–107 (2008),
   doi:http://dx.doi.org/10.1007/s10791-007-9039-3
10. Duda, R.O., Hart, P.E., Stork, D.G.: Pattern Classification. John Wiley and Sons, Inc., New York (2001)
11. Giacinto, G.: A nearest-neighbor approach to relevance feedback in content based image retrieval. In: CIVR 2007: Proceedings of the 6th ACM International Conference on Image and Video Retrieval, pp. 456–463. ACM, New York (2007),
    doi:http://doi.acm.org/10.1145/1282280.1282347

12. Giacinto, G., Roli, F.: Bayesian relevance feedback for content-based image retrieval. Pattern Recognition 37(7), 1499–1508 (2004), doi:http://dx.doi.org/10.1016/j.patcog.2004.01.005

13. Huang, T., Dagli, C., Rajaram, S., Chang, E., Mandel, M., Poliner, G., Ellis, D.: Active learning for interactive multimedia retrieval. Proceedings of the IEEE 96(4), 648–667 (2008), doi:10.1109/JPROC.2008.916364

14. Huiskes, M.J., Lew, M.S.: The MIR flickr retrieval evaluation. In: Lew, M.S., Bimbo, A.D., Bakker, E.M. (eds.) Multimedia Information Retrieval, pp. 39–43. ACM (2008), doi:http://doi.acm.org/10.1145/1460096.1460104

15. Lew, M.S., Sebe, N., Djeraba, C., Jain, R.: Content-based multimedia information retrieval: State of the art and challenges. ACM Trans. Multimedia Comput. Commun. Appl. 2(1), 1–19 (2006), doi:http://doi.acm.org/10.1145/1126004.1126005

16. Lux, M., Chatzichristofis, S.A.: Lire: lucene image retrieval: an extensible java cbir library. In: MM 2008: Proceeding of the 16th ACM International Conference on Multimedia, pp. 1085–1088. ACM, New York (2008), doi:http://doi.acm.org/10.1145/1459359.1459577

17. Rui, Y., Huang, T.S.: Relevance feedback techniques in image retrieval. In: Lew, M.S. (ed.) Principles of Visual Information Retrieval, pp. 219–258. Springer, London (2001)

18. Segarra, F.M., Leiva, L.A., Paredes, R.: A relevant image search engine with late fusion: mixing the roles of textual and visual descriptors. In: Pu, P., Pazzani, M.J., André, E., Riecken, D. (eds.) IUI, pp. 455–456. ACM (2011)

19. Smeulders, A.W.M., Worring, M., Santini, S., Gupta, A., Jain, R.: Content-based image retrieval at the end of the early years. IEEE Trans. Pattern Anal. Mach. Intell. 22(12), 1349–1380 (2000), http://www.computer.org/tpami/tp2000/i1349abs.html

20. Tamura, H., Mori, S., Yamawaki, T.: Textural features corresponding to visual perception. IEEE Trans. Systems, Man and Cybernetics 8(6), 460–473 (1978), doi:10.1109/TSMC.1978.4309999

21. Tax, D.M.: One-class classification. Ph.D. thesis, Delft University of Technology, Delft, The Netherlands (2001), doi:http://prlab.tudelft.nl/sites/default/files/thesis.pdf

22. Tong, S., Chang, E.: Support vector machine active learning for image retrieval. In: Proc. of the 9th ACM Intl Conf. on Multimedia, pp. 107–118 (2001), doi:http://doi.acm.org/10.1145/500141.500159

23. Tronci, R., Falqui, L., Piras, L., Giacinto, G.: A study on the evaluation of relevance feedback in multi-tagged image datasets. In: International Symposium on Multimedia, pp. 452–457 (2011), doi:http://doi.ieeecomputersociety.org/10.1109/ISM.2011.80

24. Wang, J.Z., Li, J., Wiederhold, G.: Simplicity: Semantics-sensitive integrated matching for picture libraries. IEEE Trans. Pattern Anal. Mach. Intell. 23(9), 947–963 (2001), doi:http://www.computer.org/tpami/tp2001/i0947abs.html

25. Zhou, X.S., Huang, T.S.: Relevance feedback in image retrieval: A comprehensive review. Multimedia Syst. 8(6), 536–544 (2003), http://www.springerlink.com/openurl.asp?genre=article&issn=0942-4962&volume=8&issue=6&spage=536

Temporal Characterization of the Requests to Wikipedia

Antonio J. Reinoso, Jesus M. Gonzalez-Barahona,
Rocio Muñoz-Mansilla, and Israel Herraiz

Abstract. This chapter presents an empirical study about the temporal patterns characterizing the requests submitted by users to Wikipedia. The study is based on the analysis of the log lines registered by the Wikimedia Foundation Squid servers after having sent the appropriate content in response to users' requests. The analysis has been conducted regarding the ten most visited editions of Wikipedia and has involved more than 14,000 million log lines corresponding to the traffic of the entire year 2009. The conducted methodology has mainly consisted in the parsing and filtering of users' requests according to the study directives. As a result, relevant information fields have been finally stored in a database for persistence and further characterization. In this way, we, first, assessed, whether the traffic to Wikipedia could serve as a reliable estimator of the overall traffic to all the Wikimedia Foundation projects. Our subsequent analysis of the temporal evolutions corresponding to the different types of requests to Wikipedia revealed interesting differences and similarities among them that can be related to the users' attention to the Encyclopedia. In addition, we have performed separated characterizations of each Wikipedia edition to compare their respective evolutions over time.

Antonio J. Reinoso · Jesus M. Gonzalez-Barahona
LibreSoft Research Group (URJC), C/ Camino del Molino s/n, 28943, Fuenlabrada,
Madrid, Spain
e-mail: ajreinoso@libresoft.es, jgb@libresoft.es

Rocío Muñoz-Mansilla
Department of Automation and Computer Science (UNED), C/ Juan del Rosal, 16,
28040, Madrid, Spain
e-mail: rmunoz@dia.uned.es

Israel Herraiz
Department of Applied Mathematics and Computing (UPM), C/ Profesor Aranguren s/n,
28040, Madrid, Spain
e-mail: israel.herraiz@upm.es

C. Lai et al. (Eds.): New Challenges in Distributed Inf. Filtering and Retrieval, SCI 439, pp. 71–89.
springerlink.com   © Springer-Verlag Berlin Heidelberg 2013

1   Introduction

Wikipedia has been an absolute revolution in the area of knowledge spreading and production. Its supporting *wiki* paradigm encourages individuals to contribute and to join efforts in the construction of a shared compendium of contents that reverts to the whole community. In this way, the on-line Encyclopedia is, nowadays, a robust and stable initiative that serves as a mass collaboration tool for building a decentralized scheme of knowledge creation. In addition, it constitutes a valuable mechanism to provide individuals access to, both, general and specialized information. The consolidation and popularity of Wikipedia is endorsed by its increasing audience that situates its web site within the six most visited ones all over the Internet[1].

Wikipedia is organized in about 270[2] editions, each one corresponding to a different language. Currently, this set of Wikipedia editions receives approximately 13,500 million visits a month, which is an absolute challenge in terms of request management and content delivery. On the other hand, Wikipedia disposes the information it offers in encyclopedic entries commonly referred as articles. At the moment of realizing this work, the different Wikipedia editions add up to almost 19 million articles and this number does not stop growing.

As a result of this relevance, Wikipedia has evolved into a subject of increasing interest for researchers[3]. In this way, quantitative examinations about its articles, authors, visits or contributions have made part of different studies [20, 11, 17]. The natural concern about the quality and reliability of the information offered has propitiated a prolific research area where several techniques and approaches have been proposed and developed [6, 4, 3, 8]. In addition, Wikipedia's growth tendencies and evolution have also been largely addressed in studies such as [2, 14, 1, 16]. Several other works have focused on particular aspects such as motivation ([7, 9]), consensus ([5, 15, 19] or vandalism [12]. By contrast, very few studies [18, 13] have been devoted to analyze the manner in which users interact and make use of Wikipedia.

Therefore, we are presenting here an empirical study consisting in a temporal characterization of users' requests to Wikipedia. Such a kind of analysis may provide different patterns involving the temporal evolution of users' interactions with the Encyclopedia and some of its results can be useful in the examination of the origin of the contributions to Wikipedia. As users of the different Wikipedias can exhibit very different habits for visiting or editing their contents, we will compare the results obtained for different editions in order to analyze the main differences and similarities among them.

---

[1] http://www.alexa.com/siteinfo/wikipedia.org (Retrieved on 22 March 2012)

[2] http://stats.wikimedia.org/EN/Sitemap.htm (Retrieved on 22 February 2012)

[3] http://en.wikipedia.org/wiki/
Wikipedia:Academic_studies_of_Wikipedia (Retrieved on 22 March 2012)

For the sake of completeness, our analysis focuses on the largest Wikipedia editions regarding their volumes of articles[4] as well as their amount of traffic[5]. In addition, we have considered a period of time corresponding to a whole year (2009) to avoid the influence of temporally localized trending effects. Thus, our main data source consists in a sample# of the users' requests submitted during 2009 to the considered Wikipedia editions. Each individual request is registered in the form of a log line by special servers deployed to deal with all the incoming traffic. Once these log lines have been sampled and sent to our facilities, they become available to be processed by an ad-hoc developed application. Basically, this application parses the log lines to extract their most relevant information fields, particularly the URL forming the request. Then, these items are filtered to determine if the corresponding log line is considered of interest for the analysis. If so, the information elements are normalized and stored in a relational database for further examination.

Our results are mainly related to the finding of patterns describing the temporal distribution of the different types of requests over time. These patterns have been studied for periods varying in length (year, months, weeks) in order to analyze periodicity and cyclical behavior at different time intervals. In respect to this subject, results involving periodical fluctuations can be particularly valuable for traffic forecasting as well as for resource re-arrangements to face periods of systems' overload. In addition, we have related our results about temporal patterns with behavioral habits exhibited by users when visiting or contributing to Wikipedia.

The rest of this chapter is structured as follows: first of all, we describe the data sources used in our analysis as well as the methodology followed to conduct our work. After this, we present our results and, finally, we present our conclusions and propose some ideas for further work.

2 The Data Sources

This section aims to describe the information sources involved in our study and used as the main data feeding to perform our analysis. Visits to Wikipedia, in a similar way to any other Web sites, are issued in the form of URLs sent from users' browsers. These URLs are registered by a set of special Squid servers which are deployed as the first layer of the Content Delivery Network of the Wikimedia Foundation and are in charge of dealing with the incoming traffic directed to all its projects.

An anonimyzed sampled of the log lines registered by the Squid layer is sent to universities and research centers interested in its study. The information contained in their different fields results of great interest as it allows to determine the nature of users' requests and, thus, to characterize them.

---

[4] `http://meta.wikimedia.org/wiki/`
`List_of_Wikipedias#Grand_Total` (Retrieved on 22 March 2012)
[5] `http://stats.wikimedia.org/EN/TablesPageViewsMonthly.htm`
(Retrieved on 22 March 2012)

Therefore, the following sections present the principal aspects related to how the Squid log lines used in this analysis are registered, their way to our storage systems and the most important information elements they contain.

2.1 The Wikimedia Foundation Squid Subsystem

Squid servers are usually used to perform web caching as proxy servers. Under this approach, they can cache contents previously browsed by users to make them available for further requests. This results in a more efficient use of network resources and in an important decrease of the bandwidth consumption. Furthermore, Squid servers may be used to speed up web servers by storing in their caches the contents repeatedly requested to them. In this way, if there are any other mechanisms, such as database or web servers, involved in the generation of dynamic contents, their participation can be avoided.

However, not all the contents asked to Wikipedia can be cached. In particular, those requested by logged users cannot be put in Squids' caches because their HTML code contain customized and per-user elements as the preferred skin. Pages requested by non-logged users are, instead, suitable of being cached as their HTML is completely similar.

Thus, when the page constituting the response to a user request can be found on a Squid server and it is up-to-date, the page is directly served from the Squid and neither the database server nor the web server have to be involved in the delivery process. Otherwise, the request is passed to the web servers which prepare the corresponding HTML code using the wiki-text stored in the database. Once the page is ready, it is submitted to the Squid layer for its caching and final delivery to the user.

The Wikimedia Foundation server architecture places, from the users' perspective, two layers of Squid servers in front of its Apache and database servers. In this way, most of the requested content is directly served from the Squid subsystem without involving the Apache servers nor the databases in the operation. Currently, there are two large Squid server clusters: a primary cluster (located in Tampa, Florida) and another secondary cluster (located in Amsterdam) that only performs web caching. The adoption of CARP (Cache Array Routing Protocol) for managing the storage and access to the cached contents allows that Squids servers run at a hit-rate of approximately 85% for text and 98% for media[6].

As the Wikimedia Foundation maintains several wiki-based projects, such us Wikipedia, Wikiversity or Wikiquote, the Squid layers have to deal with all the traffic directed to these projects. As a part of their job, Squid systems do log information about every request they serve whether the corresponding contents stem from their caches or, on the contrary, are provided by the web servers. In the end, Squid systems register a log line with different kind of information for each served request.

---

[6] http://www.nedworks.org/~mark/presentations/san/
Wikimedia%20architecture.pdf (Retrieved on 22 March 2012)

The Squid systems send the lines they register and corresponding to all the Wikimedia Foundation projects to a central aggregator host. A special program that runs there receives all the lines and sends them to a set of specific destinations. Another program samples the log streaming and packets the extracted lines to different universities or research institutions as ours. Finally, a syslog-ng client running in our facilities receives the log lines and writes them to a log file which is daily rotated.

Since each log line stored by a Squid server corresponds to a request submitted by a user and, among several other information, includes the URL summarizing its request, it constitutes a really valuable information source.

2.2 The Data Feed

Requests issued by users ask for Wikipedia's different types of contents and actions. As previously mentioned, these requests are expressed in the form of URLs and are conveniently stored once they have been satisfied. We have been provided with a sample of the requests directed not only to Wikipedia but also to all the projects maintained by the Wikimedia Foundation. The sample has been taken using a sampling factor of 1/100 during the whole year 2009 and, in general terms, it consists of more than 14,000 million requests, registered each in its corresponding log line.

Wikipedia contents are structured by means of organizational elements referred as *Namespaces*. For example, the *Main* namespace corresponds to the pages presenting the information contained in articles and is the one we solicit when we are just browsing the Encyclopedia. In addition, every article may have its corresponding *discussion page* to receive users' suggestions devoted to improve its quality. Thus, all these pages are grouped together under the *Discussion* namespace. In what our analysis is concerned, apart from the two namespaces described above, we have also considered the *User* and *User Discussion* namespaces. These namespaces aggregate, respectively, the pages arranged for each registered user and their corresponding discussion ones. In addition, we have included the *Special* namespace, which is the one gathering the pages dynamically generated in response to specific users' demands as search operations, articles linking to a particular one, and so forth.

In respect to the actions requested to Wikipedia, we will focus on visits to articles, edits, edit requests, submit and historic petitions and search operations. Visits to articles are requests just devoted to obtain their content pages. Edit operations are those intended to modify articles' information and cause the issue of write operations to the database servers. In turn, edit requests are sent when users just click on the "edit" tab placed on top of the articles' pages and lead to the reception of the corresponding *wikitext* inside a basic editor that allows to perform any changes. Submit operations are those directed to preview the results of the modifications realized on the current content of an article or to highlight the differences introduced by a given edit operation in curse. History requests present the different revisions (edit operations) performed on an article's content and leading to its actual version and state. Finally, search operations consist in requests for articles containing a given word or sentence.

3 Methodology

The study we are describing in this chapter bases on the analysis of the Wikipedia requests contained in the 1/100 sample of the global traffic to the Wikimedia Foundation projects during 2009. From all these projects, we have just focused on Wikipedia. Moreover, to ensure that the study involved mature and highly active Wikipedias, only the requests corresponding to the ten largest editions, in number of both articles and visits, have been considered. These editions are the German, English, Spanish, French, Italian, Japanese, Dutch, Polish, Portuguese and Russian ones.

Once the log lines from the Wikimedia Foundation Squid systems have been received in our facilities and conveniently stored, they become ready to be analyzed by the tool developed for this aim: The *WikiSquilter project*[7]. The analysis consists in a characterization based on a three-step process: parsing, filtering and storage. First, lines are parsed to extract relevant information elements about users' requests. Then, these information elements are filtered to determine if the corresponding requests fit the directives of the analysis. Finally, information fields from requests considered of interest are normalized and stored in a database to perform subsequent statistical examinations.

Important information about users' requests, such as the date of issue or if they caused a write operation to the database, can be extracted directly from the fields of their corresponding log lines. However, most of the data needed to perform a proper characterization are embedded in the URL representing each request that, hence, has to be specifically parsed to extract these information elements.

More in the detail, the application parser is devoted to determine the following information items:

1. The Wikimedia Foundation project, such us Wikipedia, Wiktionary or Wikiquote, to which the URL is directed.
2. The corresponding language edition of the project.
3. When the url requests an article, its namespace.
4. The action (edit, submit, history review...) requested by the user (if any).
5. If the URL corresponds to a search request, the searched topic
6. The title of every requested article or user page name.

In respect to the implementation, the parser, first, uses regular expressions to obtain the basis of the syntactical structure of the URL. Then, the different information components are extracted using common functions for string manipulation.

On the other hand, the filtering process determines which of the received log lines correspond to requests having interest for our analysis. This is accomplished

---

[7] http://sourceforge.net/projects/squilter
(Retrieved on 22 March 2012)

by checking whether the different information elements extracted from their respective URLs have been indicated to be filtered in the specific configuration file of the application. In the end, information from requests considered of interest and useful for their characterization is normalized and stored in a relational database.

Regarding the filter' implementation, it relies on an special hash structure which gathers all the elements to be filtered as well as their corresponding normalized database codes. The filter has to be queried about all the different information elements parsed from the URLs, so its accuracy and efficiency result, both, determinant for the overall application performance. Considering such great impact, important efforts have been undertaken to reduce its complexity as much as possible. As a result, filter operates with O(1) complexity because of its hash basis.

As a whole, the application has been designed and developed with strong adherence to the principles of efficiency, robustness and accuracy. Furthermore, flexibility and extensibility capabilities have been also absolutely considered. Efficiency is achieved through several elements such as multithreaded design and filter's O(1) complexity. In turn, robustness has been implemented enabling log line pre-processing to prevent inappropriate characters as well as making possible the detection of malformed URLs. As a result, absolutely all of the lines to be analyzed have been processed and none of them has caused the application malfunction. Flexibility makes the application ready to easily change the elements to be filtered by just specifying them in an XML file. This feature makes the tool able to process log information from other wiki-based projects rather than Wikipedia, again by including the items to be filtered in the aforementioned configuration file. Finally, the software architecture of the application allows to rapidly include new services and functionalities, so extensibility has been also considered.

4 Analysis and Results

In the following we are presenting our most important results about the temporal characterization of users' requests to Wikipedia. To begin with, we check if the traffic to Wikipedia can serve as an accurate model of the overall traffic to all the Wikimedia Foundation's projects. After this, we compare the evolution of the different types of requests over time. In this way, we introduce different patterns found while examining temporal behavior. In special, we will pay special attention to those showing repetitive schemes. This examination has been conducted under a comparative approach to determine whether or not the same tendencies are maintained in the considered editions of Wikipedia.

4.1 Modelling the Traffic to the Wikimedia Foundation

Figure 1 compares the traffic to the whole set of Wikipedia editions during 2009 with the overall traffic to all the projects maintained by the Wikimedia Foundation. Furthermore, it also plots the number of requests filtered after our analysis. As we can see, all three lines, each in its corresponding scale, present a relative similar

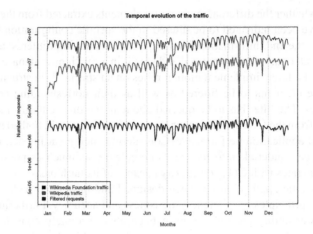

Fig. 1 Evolution of the traffic throughout 2009.

behavior over time. The decrease appreciated since November till the end of the year was due to a problem in the reception of the UDP packets sent by the Squid servers[8]. The slumps in the number of visits that appear in February, June, July and October correspond to days in which we were not able to receive and store the log lines from the Wikimedia Foundation's systems due to technical problems related to our system storage capacity.

In order to examine more accurately the relationship between the traffic to Wikipedia and to all the Wikimedia Foundation projects, Figure 2 shows the correlation between the daily measures of both traffics corresponding to the entire year. As it is shown, there is a positive correlation between the two variables so, effectively, Wikipedia traffic can serve as model for the overall traffic to the different Wikimedia Foundation projects.

4.2 Temporal Evolution of the Different Types of Requests to Wikipedia

Figures 3 and 4 show how the different types of requests to Wikipedia evolve throughout the entire year 2009. According to these figures, only those URLs involving visits, searches and edit requests would exhibit temporal repetitive patterns. On the other hand, requests consisting in edit operations, history reviews or submits for previewing contents would not present such cyclical evolutions over time. This is likely due to the fact that the requests exhibiting repetitive behaviors correspond to the most usual or generalized types of requests composing the traffic to

---

[8] More information about this issue can be found at
http://stats.wikimedia.org/EN/TablesPageViewsMonthly.htm
(Retrieved on 22 March 2012)

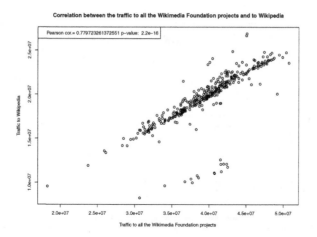

Fig. 2 Correlation between the traffic to Wikipedia and to the whole set of Wikimedia Foundation projects throughout 2009.

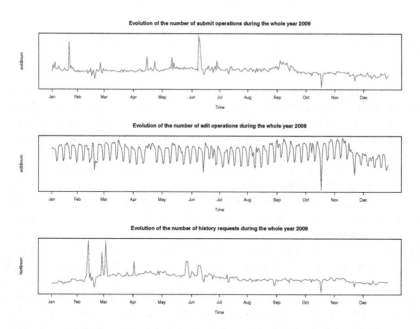

Fig. 3 Evolution of submits, edit requests and history reviews throughout 2009.

Wikipedia. Other requests, on the contrary, have a more specialized character and, thus, they appear rarely in the traffic. As a result, the most common requests follow the same periodical evolution than the general traffic to Wikipedia whereas the rest of requests show a more spurious behavior.

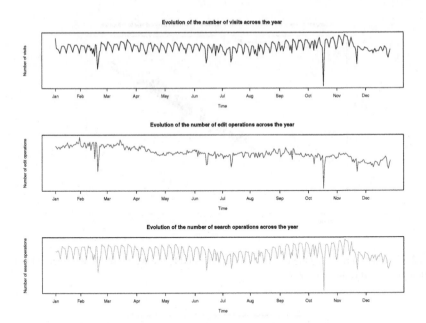

Fig. 4 Evolution of visits, edit operations and search requests throughout 2009.

In a closer approximation, let us compare the monthly evolution of visits and edits and, after, the different types of filtered actions. Edits and visits are always considered in respect to a certain Wikipedia edition because of our interests in patterns corresponding to particular communities of users. In this way, Figure 5 shows the monthly evolution of visits and edits submitted to the English and German Wikipedias [9]. Moreover, visits presented in Figure 5 correspond to articles in the *Main* namespace which is the one involved in common read operations. The idea, here, is to compare, not the figures, but the tendency during the different months analyzed and, as it can be observed, visits and edits follow considerably similar temporal evolutions.

Figures 6 and 7 present the monthly evolution of edit requests, edit operations, and history, submit and search requests for the considered Wikipedias. Although these figures are very similar in scale, we have preferred to present them using a logarithm scale in order to obtain more differentiated lines and, by means of this, a higher level of detail. As it can be observed from the chart, edit requests and searches present relatively similar evolutions as visits are not considered in this examination.

To disaggregate monthly data and to gain a more detailed information, we undertook the analysis of the different types of requests focusing on the whole weeks from year 2009. The aim was to determine whether there are patterns involving any type of requests that are repeated (periodicity) throughout the days of the week

---

[9] For the rest of considered Wikipedias:
http://gsyc.es/~ajreinoso/thesis/figures/monthVisEd.eps

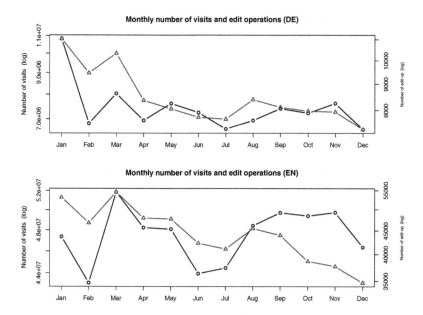

Fig. 5 Number of monthly visits to articles and monthly edit operations in the considered Wikipedias throughout 2009. The blue line reflects the visits while the red line is related to the edit operations. Left y-axis corresponds to the scale for visits whereas the right one corresponds to the scale for edit operations. In this way, values for the visits line have to be transported to the left y-axis and the ones for the edits line are in the right y-axis. The graph is presented in this way because visits and edits operations are very different in scale so presenting them together will cause a considerable loose of detail in their tendency examination.

disregarding changes in months. This is done, for example, in Figure 8 for the German, English, Spanish and French Wikipedias. More information about the rest of analyzed Wikipedia editions can be found at [10]. This newly closer perspective confirms the cyclical weekly evolution of visits, searches and edit requests. On the contrary, it is much more difficult to pronounce about the periodicity of the rest of actions (specially edits) because of their more varying character and their lower number of requests. In addition, the temporal distributions of requests may substantially vary depending on each edition of Wikipedia. As an example, Figure 9 presents the charts corresponding to the Spanish and Japanese Wikipedias. The former presents relatively well-defined and identifiable patterns whereas the latter shows more irregular distributions. In general, all the editions present a weekly repetitive pattern for visits except the Japanese and, perhaps, the Polish Wikipedias which do not show such well defined patterns. The temporal distributions of requests in these two editions of Wikipedia absolutely deserves further examination as the differences they exhibit may be related to particular sociological or

---

[10] http://gsyc.es/~ajreinoso/thesis/

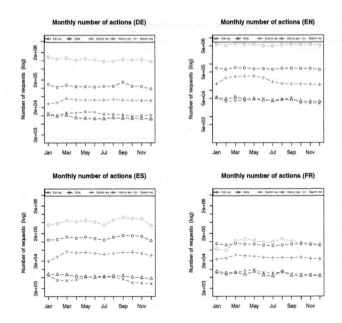

Fig. 6 Monthly distribution of the different types of actions in different Wikipedias.

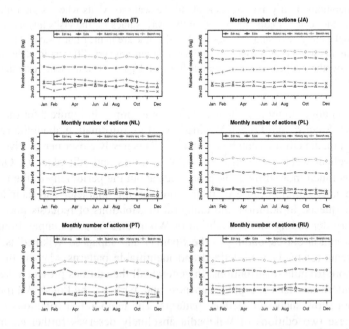

Fig. 7 Monthly distribution of the different types of actions in different Wikipedias.

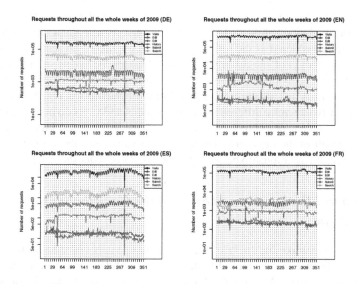

Fig. 8 Evolution of the different types of requests during every whole week of 2009 (DE EN ES FR).

sociocultural aspects or to the fact that their respective communities of uses are within the most geographically enclosed.

Stationarity can be assessed using the autocorrelation function (ACF). In this way, Figure 10 shows the autocorrelation function of the visits and edit operations as well as of the edit, history, submit and search requests in the English Wikipedia As Figure 10 shows, visits, searches and requests for editing exhibit clear periodicities corresponding to the correlation between values separated by 7 units, i.e., between weekly values. In turns, edit and submit or history requests do not present such well-defined cyclic behavior, although they present a certain stationary evolution also considering periods of 7 days. Thus, after the autocorrelation analysis, all of the considered Wikipedia present easily appreciable periodicities except the Japanese one.

As we are considering only whole weeks, we can merge the requests corresponding to each day of the week in order to obtain a unified picture of the overall behavior as the week advances. In this way, if we aggregate the different types of requests and analyze their distributions over the days of the week, as presented in Figure 11 for the German Wikipedia, we can appreciate that some of the requests, specifically visits, searches and requests for editing, are similarly distributed throughout the days of the week in all the considered Wikipedias. Edits, history and submit requests, however, present more remarkable differences among the different editions and, consequently, they adopt more different patterns. Nevertheless, in the case of the German, English, Spanish, Italian and Russian Wikipedias edits conserve a relatively similar shape that also match the evolution of visits.

This subject can be further examined using the cross-correlation function (CCF) to compare the evolutions of the different types of requests with the temporal distribution of visits, considered as the reference element. Figure 12 presents the results of the cross-correlation of the different types of requests and visits in the English Wikipedia. According to this Figure, requests for editing and searches follow similar evolutions as visits. Edits also present a quite similar behavior whereas history and submit requests evolve more differently. However, edits and visits do not present the same similarity in their respective evolutions in all the Wikipedias. Figure 12 also includes the result of the cross-correlation between edits and submit requests that indicates a temporal relationship between the two types of requests.

4.3   Temporal Evolution and Origin of the Contributions to Wikipedia

We decided to undertake the study of the evolution of visits and edits at the level of the days of the week in the aim of finding a meaningful closeness between their two temporal variations. As a result of such kind of analysis, Figure 13 presents the evolution of both types of requests throughout the days of the week for all the considered Wikipedias. Visits and edits, in each Wikipedia edition, correspond to the overall year and have been grouped by their day of issue. So, Figure 13 presents their compared progressions and shows a considerably closeness in the evolution of

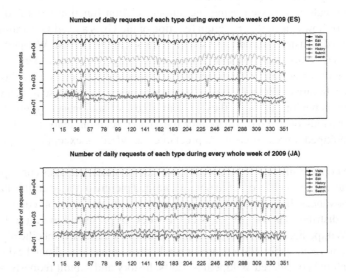

Fig. 9 Number of daily requests of each different type issued for every whole week of 2009. This chart presents the evolution of each kind of request during every whole week of 2009 in different editions of Wikipedia. X-axis begins with the first Monday of the year and finishes with the last Sunday and each vertical pair of divisions delimit an entire week.

both types of requests in several Wikipedias. Nevertheless, the number of edits tends to raise in weekends for a group of them (French, Japanese, Dutch and Polish). That could mean that, in those editions, editors are not part of the great mass of people visiting the articles but just a minor group devoted to contribute or to maintain them.

Moreover, Figure 14 presents the weekly distributions of edits and edit requests. Again, we have to pay attention to the different edges and scales for each type of action. In other case, it would seem that there are more edit operations on some days (specially on Saturdays) than requests for editing (impossible situation because every edit operation has to be preceded by the corresponding edit request). The graph shows how requests for editing and completed edits are closer on Saturdays than in any other week day for some of the considered Wikipedias. This is due to the fact that on Saturdays in the French, Japanese, Polish and Dutch Wikipedias, edit requests decrease whereas finished edit operations raise. In other words, almost every edit request submitted on Saturday in these Wikipedias ends with the corresponding write operation to the database. This can be seen as a reinforcement of the existence of a group of more productive editors in these Wikipedias during weekends.

We have compared our results about distribution of authoring with the ones obtained by Ortega in [10]. Ortega used Gini coefficients to determine the degree of the concentration of edits over the communities of authors corresponding to the different editions of Wikipedia. High values of these coefficients would mean high concentration of edits and, thus, a reduced community of effective authors. In this way, we found that editions with the highest Gini coefficient according to Ortega (Dutch, Portuguese and French) are within the ones we consider as having an elite of authors because of their distributions of visits and edits.

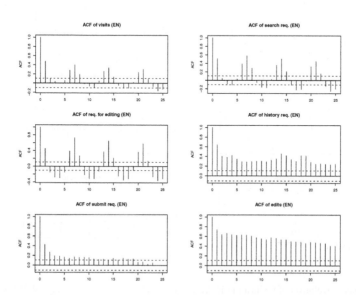

Fig. 10 Auto-correlation of the different types of requests in the English Wikipedia throughout 2009.

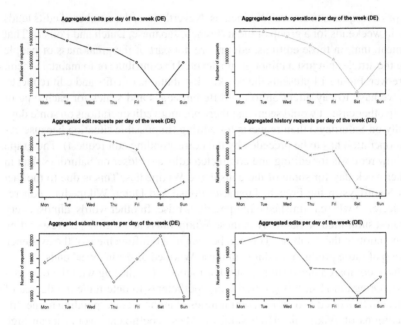

Fig. 11 Evolution of the different types of requests throughout the days of the week (DE).

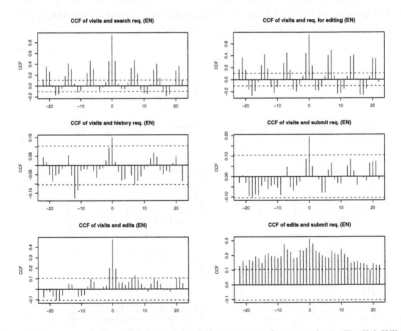

Fig. 12 Cross-correlation of visits and the different types of requests in the English Wikipedia throughout 2009 (Cross-correlation between edits and submit requests is also included)

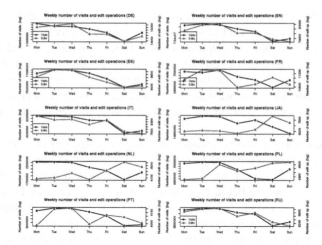

Fig. 13 Evolution of visits and edits throughout the days of the week in the different editions of Wikipedia.

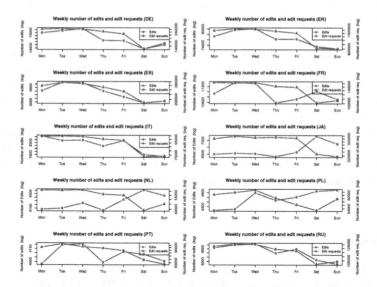

Fig. 14 Evolution of edits and submit requests throughout the days of the week in the different editions of Wikipedia.

5 Conclusions and Further Work

We can extract several conclusions after our efforts for characterizing temporarily the requests submitted to Wikipedia. First of all, we have shown how temporal information related to users' requests can be obtained from log lines stored by Wikimedia

Foundation's Squid servers. Using this information we have modeled the variations over time of the different kind of requests submitted by users to Wikipedia. Our first finding was the fact that requests to Wikipedia temporarily model the overall traffic to all the Wikimedia Foundation projects. Of course it was what we were expecting, as Wikipedia is, by far, the most trending project maintained by the Wikimedia Foundation. However, we managed to obtain a high degree of correlation between Wikipedia's traffic and the requests directed to all the Wikimedia Foundation projects.

We have also illustrated how demands to Wikipedia consisting in visits, searches and edit requests present repeated patterns over time as they are the most generally solicited. On the other hand, submit or history requests and edits present a spurious and irregular nature because of their most specific character. In relation to this topic, the size of the sample may be determinant as the low percentage of edits contained in it can prevent the observation of cyclical distribution. So, further examinations should involve higher sampling factors to accurately analyze the presence of stationarity in the distribution over time of edits. Nevertheless, we have used the auto-correlation and cross-correlation functions in an effort to determine both the existence of periodicity in the temporal distributions of users' requests as well as the nature of the relationship between requests consisting in visits and the rest of the different actions considered in our analysis.

We have also related the temporal distribution of the different types of requests to Wikipedia with behavioral aspects related to its users. In this way, we have exposed how, in some editions, general visitors tend to be more participative and, at a given time, they turn into editors. In other Wikipedias, otherwise, it seems that a purportedly elite of authors contribute to the Encyclopedia during its spare time, often during weekends.

In the future, we plan to add geolocation to the temporal characterization process. In this way, a reference time plus the geographical position could better serve to determine the habits of the different communities of users when browsing Wikipedia. Furthermore, a closer analysis of the evolution of the different types of requests will allow to find more accurately defined relationships among them.

References

1. Buriol, L.S., Castillo, C., Donato, D., Leonardi, S., Millozzi, S.: Temporal analysis of the wikigraph. In: Proceedings of the 2006 IEEE/WIC/ACM International Conference on Web Intelligence, WI 2006, pp. 45–51. IEEE Computer Society, Washington, DC (2006)
2. Capocci, A., Servedio, V.D.P., Colaiori, F., Buriol, L.S., Donato, D., Leonardi, S., Caldarelli, G.: Preferential attachment in the growth of social networks: the case of wikipedia (February 2006)
3. Chesney, T.: An empirical examination of wikipedia's credibility. First Monday 11(11) (November 2006)
4. Giles, J.: Internet encyclopaedias go head to head. Nature 438(7070), 900–901 (2005)

5. Kittur, A., Suh, B., Pendleton, B.A., Chi, E.H.: He says, she says: conflict and coordination in wikipedia. In: CHI 2007: Proceedings of the SIGCHI Conference on Human Factors in Computing Systems, pp. 453–462. ACM Press, New York (2007)
6. Korfiatis, Nikolaos, Poulos, Marios, Bokos, George: Evaluating authoritative sources using social networks: an insight from wikipedia. Online Information Review 30(3), 252–262 (2006)
7. Kuznetsov, S.: Motivations of contributors to wikipedia. SIGCAS Comput. Soc. 36(2) (June 2006)
8. Nielsen, F.A.: Scientific citations in wikipedia (May 2007)
9. Nov, O.: What motivates wikipedians? Commun. ACM 50(11), 60–64 (2007)
10. Ortega, F.: Wikipedia: A quantitative analysis. PhD thesis, Universidad Rey Juan Carlos (2009), http://libresoft.es/Members/jfelipe/phd-thesis
11. Ortega, F., Gonzalez-Barahona, J.M., Robles, G.: The top ten wikipedias: A quantitative analysis using wikixray. In: Proceedings of the 2nd International Conference on Software and Data Technologies (ICSOFT 2007). INSTICC. Springer (July 2007)
12. Priedhorsky, R., Chen, J., Lam, S.K., Panciera, K., Terveen, L., Riedl, J.: Creating, destroying, and restoring value in wikipedia. MISSING (November 2007)
13. Reinoso, A.J.: Temporal and behavioral patterns in the use of Wikipedia. PhD thesis, Universidad Rey Juan Carlos (2011), http://gsyc.es/~ajreinoso/phdthesis
14. Spinellis, D., Louridas, P.: The collaborative organization of knowledge. Commun. ACM 51(8), 68–73 (2008)
15. Suh, B., Chi, E.H., Pendleton, B.A., Kittur, A.: Us vs. them: Understanding social dynamics in wikipedia with revert graph visualizations. In: 2007 IEEE Symposium on Visual Analytics Science and Technology, pp. 163–170. IEEE (October 2007)
16. Suh, B., Convertino, G., Chi, E.H., Pirolli, P.: The singularity is not near: slowing growth of wikipedia. In: WikiSym 2009: Proceedings of the 5th International Symposium on Wikis and Open Collaboration, pp. 1–10. ACM, New York (2009)
17. Tony, S., Riedl, J.: Is wikipedia growing a longer tail? In: GROUP 2009: Proceedings of the ACM 2009 International Conference on Supporting Group Work, pp. 105–114. ACM, New York (2009)
18. Urdaneta, G., Pierre, G., van Steen, M.: A decentralized wiki enginge for collaborative wikipedia hosting. In: Proceedings of the 3rd International Conference on Web Information Systems and Technologies, pp. 156–163 (March 2007)
19. Viégas, F.B., Wattenberg, M., Kriss, J., van Ham, F.: Talk before you type: Coordination in wikipedia. In: MISSING, p. 78 (2007)
20. Voss, J.: Measuring wikipedia. In: 10th International Conference of the International Society for Scientometrics and Informetrics, ISSI (July 2005)

Interaction Mining: The New Frontier
of Customer Interaction Analytics

Vincenzo Pallotta and Rodolfo Delmonte

Abstract. In this paper, we present our solution for argumentative analysis of call center conversations in order to provide useful insights for enhancing Customer Interaction Analytics to a level that will enable more qualitative metrics and *key performance indicators* (KPIs) beyond the standard approach used in Customer Interaction Analytics. These metrics rely on understanding the dynamics of conversations by highlighting the way participants discuss about topics. By doing that we can detect relevant situations such as social behaviors, controversial topics, customer oriented behaviors, and also predict customer satisfaction.

1 Introduction

Call centers data represent a valuable asset for companies, but it is often underexploited for business purposes. By call center data we mean all information that can be gathered from recording calls between representatives (or agents) and customers during their interactions in call centers. These interactions can happen over multiple different channels including telephone, instant messaging, email, web forms, etc. Some information can be collected without looking at the content of the interaction, by simply logging the system used for carrying the conversation. For example, in call centers, calls duration or number of handled calls can be measured by software for telephony communication. We call these measures standard call center Key Performance Indicators (KPIs). With standard KPIs, only

Vincenzo Pallotta
University of Business and International Studies
Av. Blanc, 46 1202 Geneva, Switzerland
e-mail: vincenzo.pallotta@gmail.com

Rodolfo Delmonte
Department of Computational Linguistics
University 'Ca Foscari Venice, Italy
e-mail: delmont@unive.it

C. Lai et al. (Eds.): New Challenges in Distributed Inf. Filtering and Retrieval, SCI 439, pp. 91–111.
springerlink.com   © Springer-Verlag Berlin Heidelberg 2013

limited analytics can be done. Of course, one can aggregate these measures over several dimensions of other meta-data such as agents, customers, regions, time, and queues. However, this only provides a partial understanding of the call center performance and no information whatsoever is collected about what is going on within the interaction.

Customer Interaction Analytics is aimed at solving the above issue by enabling tapping into the content of conversations. The technology for Customer Interaction Analytics is still in its infancy and related commercial products have not yet achieved maturity. This is due to two main factors: i) it is highly dependent on quality of speech recognition technology and ii) it is mostly based on text-based content analysis.

We believe that text-based content analysis approaches are highly sensitive to input quality and that conversational input is fundamentally different than text. Therefore, conversations should be treated differently. Natural Language Processing (NLP) technology needs be adapted to and robust enough to deal with the conversational domain in order to achieve acceptable performance. Moreover, the level of analysis of conversation cannot be set to semantics only. It must consider the purpose of language in its context, i.e., pragmatics.

Our approach to Customer Interaction Analytics is based on Interaction Mining. Interaction Mining is a new research field aimed at extracting useful information from conversations. In contrast to Text Mining (Feldman and Sanger 2006), Interaction Mining is more robust, tailored for the conversational domain, and slanted towards *pragmatic* and *discourse* analysis. We applied our technology for pragmatic analysis of natural language to a corpus of call center conversations and shown how this analysis can deal with the situations we mentioned earlier. In particular, with our approach we were able to achieve the following objectives:

1. Identify Customer Satisfaction in call center conversations. As shown by Rafaeli et al. (2007), this metrics is predicted by so called "customer oriented behaviors";
2. Identify Root Cause of Problems by looking at controversial topics and how agents are able to deal with them;
3. Identify customers who need particular attention based on history of problematic interactions;
4. Learn best practices in dealing with customers by identifying agents able to carry cooperative conversations. This knowledge coupled with customer profiles can be used effectively in intelligent skill-based routing[1]

The article is organized as follows: in section 2 we review current Speech Analytics technology and make the case for Interaction Mining approach in order to address the current business challenges in call centers quality monitoring and assessment. In section 3 we present our Interaction Mining solution based on a specific kind of pragmatic analysis: the argumentative analysis and its implementation with the A3 algorithm. In section, 4 we present the four business cases outlined above, by showing the analysis of actual call center data and the

---

[1] http://en.wikipedia.org/wiki/Skills-based_routing

implementation of new relevant metrics and KPIs for call center quality monitoring. We conclude the article with a discussion on the achieved results and a roadmap for future work.

2 Customer Interaction Analytics Needs Interaction Mining

Call center data contain a wealth of information that usually remains hidden. Key Performance Indicators (KPIs) for call centers performance can be classified into three broad categories (Baird 2004):

1. Agent Performance Statistics: these include metrics such as *Average Speed of Answer, Average Hold Time, Call Abandonment Rate, Attained Service Level,* and *Average Talk Time.* They are based on quantitative measurements that can be obtained directly through ACD[2] Switch Output and Network Usage Data.
2. Peripheral Performance Data: these include metrics such as *Cost Per Call, First-Call Resolution Rate, Customer Satisfaction, Account Retention, Staff Turnover, Actual vs. Budgeted Costs,* and *Employee Loyalty.* These metrics are mostly quantitative, with the exception of *Customer Satisfaction* that is usually obtained through Customer Surveys. The quantitative metrics are usually collected through Sales Records Expense Records, Human Resources Service Records, and Financial Records.
3. Performance Observation: these include metrics such as *Call Quality, Accuracy and Efficiency, Adherence to Script, Communication Etiquette,* and *Corporate Image Exemplification.* These are qualitative metrics based on analysis of recorded calls and session monitoring by a supervisor.

Minnucci (2004) reports that the most required metrics by call center managers are indeed the qualitative ones topped by Call Quality (100%) and Customer Satisfaction (78%). However, these metrics are difficult to implement with the adequate level of accuracy[3]. For instance, the Baird study (2004) points out that for Customer Satisfaction, accuracy can be:

> "negatively affected by insufficient number of administered surveys per agent resulting in not enough samples of individual agent's work to constitute a representative sample. The result could be an unfair judgment of the agent's performance and allocations of bonuses based more upon chance, good fortune than merit."

This problem would disappear if a more systematic analysis would be conducted over the entire corpus of recorded calls with no human intervention on observation. Therefore, turning such a qualitative metric into a quantitative one is certainly an important challenge that could move Customer Interaction Analytics a leap forward.

---

[2] Automatic Call Distribution.
[3] Accuracy is defined in (Baird 2004) as true indication and it depends on the actual level of performance attainment, especially with regard to statistical validity.

Most call center quality monitoring dashboards are now only able to display information related to service-level measures (Agents and Peripheral Performance data), namely how fast and how many calls agents able to handle. Fig. 1 is an example of such a dashboard.

Because of recent improvements of Automatic Speech Recognition (ASR) technology (Neustein 2010), *Speech Analytics* is viewed as a key element implementing call center quality monitoring where ASR technology is leveraged to implement relevant KPIs. As pointed out by Gavalda and Schlueter (2010), Speech Analytics is becoming

> "an indispensable tool to understand what is the driving call volume and what factors are
> affecting agents' rate of performance in the contact center."

However, current Speech Analytics solutions have focused on *search* rather than *information extraction*. Most system allows the user to search the occurrences of keywords or key-phrases in the spoken conversations (i.e. audio files). While this represents an important feature for targeted call monitoring, it fails in delivering an understanding of the context where such terms occur. In other words, this method can be helpful when the supervisor has a clear idea of what to look for in recorded conversation but it helps very little in data mining and intelligence.

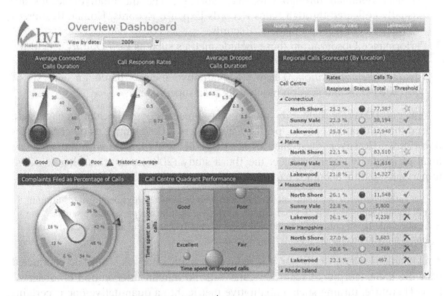

Fig. 1 An example of Call Center dashboard[4] implementing standard metrics

---

[4] An example of Customer Interaction Analytics dashboard is available at
http://demos7.dundas.com/HVR.aspx

2.1 Interaction Mining

Interaction Mining is an emerging field in Business Analytics that contrasts the standard approach based on Text Mining (Feldman and Sanger 2006). In Text Mining the assumption made is that input is textual and can be treated just as a container of semantic content where non-content words can be filtered out. This assumption is no longer valid in conversational input. Non-content words such as conjunctions, prepositions, personal pronouns and interjections are extremely important in conversations and cannot be filtered out as they bear most of their *pragmatic meaning*. Hence Text Mining tools cannot be entirely transferred to Interaction Mining. Some of the Text Mining tools are still useful such as Entity Recognition, Tokenization and Part-of-Speech Tagging. As pointed out in Pallotta et al. (2011) there are several advantages of moving to Interaction Mining for generating intelligence from conversational content.

It is important to note that while the purpose is similar – i.e., turning unstructured data into structured data for performing quantitative analysis - Text Mining focuses on pattern extraction from *documents*. This is justified because of the massive presence of content words in textual input (e.g. news, articles, blogs, corporate documents). As we mentioned earlier, this is no longer the case with conversational content. The units of information in conversational content are *dialogue turns* and typically they are significantly shorter than documents considered as input for Text Mining. This means that the input has to be linguistically processed in order to understand its *pragmatic function* in the conversation. For instance, a simple turn containing just one single word like "Yes" or "No" can make a substantial difference in the interpretation of a whole conversation. In other words, in Text Mining, the conversational meaning and the context are likely to get lost, which is instead kept and taken into account in Interaction Mining.

Interaction Mining tools are substantially different than those employed in Text Mining. First of all, statistical or extensive machine learning approaches are no longer a viable option since data are very sparse. While it is possible to learn patterns from content-bearing documents, it is nearly impossible to learn pragmatic meaning from non-content bearing words. Approaches that attempted to apply machine learning to Interaction Mining have failed in providing satisfactory results so far (Rienks and Verbree 2006; Hakkani-Tür 2009). This can be explained because the amount of data needed for supervised learning becomes an insurmountable bottleneck. The requirements for Interaction Mining tools are that conversational units, the turns, have to be interpreted in their linguistic context. It is simply not possible to consider them as isolated input, as it is the case with documents or web pages. Moreover, accuracy requirements are higher than those required for Text Mining. For instance, in text categorization, content-words can be used for discrimination and highly frequent non-content words can be removed. This is not the case in conversations where highly frequent non-content words cannot be removed as they carry pragmatic meaning (e.g. prepositions, conjunctions). In Pallotta et al. (2011) we have provided evidences that bag-of-words approach simply is not suitable for pragmatic indexing of conversations, and therefore useless for tasks as Question Answering or Summarization.

Another limitation of Text Mining approach to conversations is in Sentiment Analysis. As Delmonte and Pallotta (2011) previously showed, shallow linguistic processing and machine learning often provide misleading results. Therefore, we advocated for a deep linguistic understanding of input data even for standalone contributions such as product reviews. In Interaction Mining, the Sentiment Analysis issues become even more compelling because sentiment about a topic is not fully condensed in a single turn but it develops along the whole conversation. For example, it is very common that dissent is expressed toward the opinion of other speakers rather than to the topic under discussion. Sentences like: "why do you think product X is bad?" would be simply mistakenly classified as carrying a negative attitude to product X in a bag-of-word approach.

2.2 Related Work

Current approaches to Customer Interaction Analytics are mostly based on Speech Analytics and Text Mining, which is essentially Search and Sentiment Analysis. Recorded speech is first indexed and searched against a set of negative terms and relevant topics. There are currently two main approaches for speech indexing: i) phonetic transcription and ii) Large Vocabulary Conversational Speech Recognition (LVCSR).

In phonetic transcription, speech recognition is made against a small set of phonemes. Keyword queries are algorithmically converted into their phonetic representation and used for search the phonetic index. With this approach one can search for occurrence of specific terms in calls. Its simplicity is at the same time its strength and weakness. On the one hand the method is fast and accurate but, on the other hand, it is limited to its applicability for generating adequate insights on calls because the context of word's occurrence is lost and it can only recovered by physically listening to the audio excerpt where the searched word occurs. One possible workaround is to systematically search for the occurrence of a large number of terms taken from a pre-defined list, thus obtaining a partially transcribed speech. The words list is typically generated by harvesting domain-related words from corporate documents or from relevant web search results. Several companies put this method forward as the most viable solution for content-based Speech Analytics (see Gavalda and Schlueter (2010) for a detailed coverage).

LVCSR is instead based on the recognition of a very large vocabulary of words and thus provide standard textual transcription of the calls. Transcription errors are usually caused by out-of-vocabulary words and measured as Word Error Rate (WER) [5]. While still considerably high compared to human performance in transcription, accuracy of LVCSR systems show a promising trend as reported by the NIST Speech-To-Text Benchmark Test History 1988-2007 (Fiscus et al. 2008).

Another common approach to the analysis of call center data is that of automatic call categorization through supervised machine learning (Gilman et al. 2004; Zweig et al. 2006; Takeuchi et al. 2009). These methods have failed in providing satisfactory results even for very broad categories. The problem still lies on data

---

[5] http://en.wikipedia.org/wiki/Word_error_rate.

sparseness and that huge amount of training data is necessary to achieve reasonable discriminatory power. Getting huge training data is not an option also because training is highly influenced by domain specificity. Transferring trained models from a domain to another would be problematic.

Unsupervised learning provides better results for domain-specific classes as shown in Tang et al. (2003). However, the sensitivity to the domain represents a big issue. Moreover, this type of categorization – i.e. topics of calls – helps little to understand if a call is satisfactory or not. It might be better suited for retrieval and aggregation of other quality-oriented information.

Instead of downgrading the analysis capabilities we believe it is more appropriate to make the analysis less sensitive to WER and domain-specificity. In other words, we want a robust solution capable of delivering approximate but still sound measurements, which are relevant for implementing call-center metrics. We will show in the next sections that our approach to Interaction Mining is robust and it can properly deal with output from LVCSR systems.

3 Argumentative Analysis for Interaction Mining

Our approach to pragmatic analysis for Interaction Mining is rooted on *argumentative analysis* (Pallotta 2006). Argumentation is a pervasive pragmatic phenomenon in conversations. Purposeful conversations are very often aimed at reaching a consensus for a decision or to negotiate opinions about relevant topics. Both types of conversations contain argumentative actions that can be recognized by Interaction Mining systems. Recognizing the argumentative structure of a conversation is useful for several tasks such as Question Answering, Summarization and Business Analytics (Pallotta et al. 2011). In the specific case of Customer Interaction Analytics, we used argumentative analysis as the basis to perform analysis of conversations and synthesize a cooperativeness score for each of them, which in turn it allows us to predict customer satisfaction for these calls.

In this section we provide a few insights on automatic argumentative analysis performed through the tailoring of an advanced Natural Language Understanding (NLU) technology. Further details are available in Pallotta and Delmonte (2011) and Delmonte et al. (2010).

3.1 Argumentative Structure of Conversations

The argumentative structure defines the various patterns of argumentation used by participants in the dialog, as well as their organization and synchronization in the discussion. A dialog is decomposed into several stages such as issues, proposals, and positions, each stage being possibly related to specific aggregations of elementary dialog acts. Moreover, argumentative interactions may be viewed as specific parts of the discussion where several dialog acts are combined to build such an interaction; for instance, a disagreement could be seen as an aggregation of several acts of reject and accept of the same proposal. From this perspective, we adopted an argumentative coding scheme, the Meeting Description Schema

(MDS), developed in Pallotta (2006). In MDS, the argumentative structure of a meeting is composed of a set of topic discussion episodes, where an episode is a discussion about a specific topic. In each discussing topic, there exists a set of episodes in which several issues are discussed. An issue is generally a local problem in a larger topic to be discussed and solved. Participants propose alternatives, solutions, opinions, ideas, etc. in order to achieve a satisfactory decision. Proposal can be accepted or challenged through acts rejecting or asking questions. Hence, for each issue, there is a corresponding set of proposals episodes (solutions, alternatives, ideas, etc.) that are linked to a certain number of related positions episodes (for example a rejection to a proposed alternative in a discussing issue) or questions and answers.

We illustrate this approach by contrasting the limitation of classical term-based indexing for retrieving relevant content of a conversation. Consider the conversation excerpt in Fig. 2 (a) and the query: *"Why was the proposal on microphones rejected?"*. A classical indexing schema would retrieve the first turn from David and by matching the relevant query term "microphone". There is no presence of other query terms such as "reject", "proposal". Moreover, it is not possible to map the "Why" question onto some query term (e.g. reason, motivation, justification, explanation). This makes impossible to adequately answer this query without any additional metadata that highlight the role of the participants' contributions in the conversation.

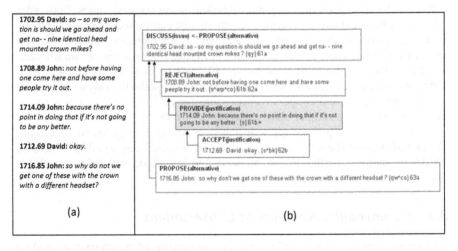

Fig. 2 Argumentative Structure of a conversation

In Fig. 2 (b), we have computed the argumentative structure of the conversation excerpt that allows us to correctly answer the question by selecting the third turn. In fact, the "Why" question is mapped to a query term, which is found as an argumentative index, "justification", for that turn. Of course, finding justification is not enough, but the retrieval algorithm needs to check whether that justification has been provided as a rejection of a "proposal" (or "alternative") made to an issue on the topic of microphones. This can only be done by navigating back in the argumentative chain up to the turn tagged as "issue" and whose content matches the term "microphone".

3.2 Automatic Argumentative Annotation

The core of our solution is a system that automatically extracts the argumentative structure of conversations, as the one shown in Fig. 2. This system is based on specific tailoring and extension of GETARUNS (Delmonte 2007; 2009), a system for text understanding developed at the University of Venice. *Automatic Argumentative Annotation* (A3) is carried out by a special module of the GETARUNS system activated at the very end of the computation of each dialog. This module takes as input the complete semantic representation produced by the system. To produce Argumentative annotation, the system uses the following 21 Discourse Relations labels: *statement, narration, adverse, result, cause, motivation, explanation, question, hypothesis, elaboration, permission, inception, circumstance, obligation, evaluation, agreement, contrast, evidence, hypothesis, setting, prohibition.*

These are then mapped onto five general argumentative labels:

1. ACCEPT,
2. REJECT/DISAGREE
3. PROPOSE/SUGGEST
4. EXPLAIN/JUSTIFY
5. REQUEST.

The 5 general argumentative categories are broke down into 12 finer grained categories such as "accept_explanation", "accept_suggestion", "provide_explanation", "reject_explanation", "reject_suggestion", etc. as described in the MDS coding schema.

Details of the A3 algorithm are available in Pallotta and Delmonte (2011) and in Delmonte et al. (2010). The system has been evaluated on conversations from the ICSI meeting corpus (Janin et al. 2001) and annotated by Pallotta et al. (2007). On a total of 2304 turns, 2251 have received an argumentative automatic classification, with a Recall of 97.53%. We computed Precision as the ratio between Correct Argumentative Labels/ Argumentative Labels Found, which corresponds to 81.26%. The F-score is 88.65%.

3.3 Robustness of the A3 Algorithm for Speech Input

The A3 algorithm was evaluated against a corpus of manually transcribed conversations. In order to test if it achieves comparable results on automatically transcribed conversations, we conducted an experiment on similar meetings that have been transcribed using state-of-the-art LVCSR technology (Fiscus et al. 2008). We have measured the performance of our system and observed an overall degradation of only 11.7% on automatically transcribed conversations with a system showing an average WER of 30% (Hain et al. 2009). This result has been obtained without any intervention on the A3 algorithm itself, which was designed initially to deal with manually transcribed data. These results are quite promising and, coupled with expected improvements in LVCSR technology and further tuning of the system, they provide us with a solid basis for development.

3.4 Multi-word Expressions

One key issue with conversation is that topics are not expressed by single words but very often by compounds. Hence, quality of topic detection can be improved if the lexicon contains domain-specific multi-word expressions. We thus run a multi-word expressions extraction tool (Seretan and Wehrli. 2009) to identify the most frequent collocations in the corpus and compare them with the topics detected by the GETARUNS system. The top 10 extracted multi-word expressions[6] (a) and topics (b) are shown in Table 1.

Table 1 Multi Word Expression extracted from the corpus (a) and GETARUN topics (b).

Multi Word Expression	Score		Topic	% of total
1. Calling Chase	475.4809		1. Chase	5.26%
2. Account number	300.2746		2. Social security number	3.41%
3. Gross balance	282.5876		3. Checking account	2.75%
4. Direct deposit	247.4588		4. Moment	2.33%
5. Savings account	189.3173		5. Statement	2.16%
6. And available	186.6647		6. Money	2.00%
7. Social security number	159.8058		7. Savings account	1.45%
8. Area code	146.8807		8. Dollar	1.37%
9. Daytime phone number	143.3286		9. Days	1.35%
10. Most recent	126.4333		10. Phone number	1.23%

a   b

While there is a predictable overlapping it is interesting to see that some domain-dependent terms were detected by the multi-word extraction system but they were not included in the lexicon of our system such as "gross balance" and "and available", and "direct deposit"[7]. We hypothesize that enriching the lexicon with these terms would greatly improve the pragmatic analysis of conversations. In our future

---

[6] The score for multi-word expression represents the log-likelihood ratio statistics representing the association strength between the component words (Dunning 1993).

[7] The "and" and "Available" are detected as a multi-word expression because they occur frequently in the corpus as the pattern "Gross and Available balance".

work, we will investigate a way to systematically include multi-word expression information in our system.

4 Business Cases with Call Center Data

In this section we describe how the output of the A3 algorithm can be used to perform Interaction Mining and we present the results of an experiment where we applied it to actual call center data. The main goal was to find out if the argumentative analysis, coupled with other standard text mining analysis (i.e. Sentiment and Subjectivity analysis), could indeed provide useful information to implement several Customer Interaction Analytics metrics. The results show, in particular, that we were able to achieve the objectives we introduced in Section 1. As already discussed in Section 2, being able to extract the above information enables us to implement the relevant and most requested KPIs in call center quality management.

4.1 The Corpus

In our experiment we used a corpus of 213 manually transcribed conversations of a help desk call center in the banking domain. Each conversation has an average of 66 turns and an average of 1.6 calls per agent. This corpus was collected and annotated for a study aimed at identifying conversational behaviors that could favor satisfactory interaction with customers (Rafaeli et al. 2007). This study has shown that is the case and that *customer-oriented behaviors* (COBs) can indeed be used to predict customer ratings. Table 2 contains the identified COBs and their distribution in the annotated portion of the dataset.

Table 2 Customer-Oriented Behaviors from Call Center data

Customer Oriented Behaviors	
Anticipating Customers Requests	22,45%
Educating The Customer	16,91%
Offering Emotional Support	21,57%
Offering Explanations / Justifications	28,57%
Personalization Of Information	10,50%

Notice that only a very small portion of the dataset was manually annotated with COBs, representing only 2.5% of the entire corpus. This prevented us to perform a statistically sound correlation study, as we cannot consider the non-annotated data as examples of conversations not containing COBs. Additionally, it was not possible to assume that turns that not received COB annotations were actually negative examples, i.e. they did not contain COBs. They were simply not annotated. This prevented us to use the dataset with machine learning algorithms.

4.2 Argumentative Analysis of Call Center Conversations

Despite the above limitations, we carried out an experiment by running the A3 algorithm on the COB annotated portion of the dataset. Examples of COB and argumentative annotations of the corpus are shown in Table 3. The table contains turns from a single conversation, which have been annotated as COBs and that are also automatically classified as relevant argumentative categories.

In the example, we show only those turns whose COB and argumentative annotations can be considered as "compatible".

Table 3 Examples of COBs from a conversation and their classification as argumentative categories.

Customer Oriented Behaviors	Argumentative Categories		
	Accept explanation	Provide explanation	Suggest
Anticipating Customers Requests			3
Cause if you like I can help you do a transfer by using our touchtone service.			1
Cause with the ATM card I can help you do a transfer by using your touchtone.			1
			1
I can send you a form that you can fill out and return back to us.			
Offering Emotional Support	1		
Thanks for calling Chase and enjoy your day Miss Asher.	1		
Offering Explanations / Justifications		1	
And that application will be out in the mail within two business days.		1	

We noticed that COBs showed a high resemblance to our argumentative categories and that they might correlate as well. The chart of Fig. 3 shows the number of COB-annotated turns (y-axis) with their argumentative labels assigned by the A3 algorithm, which classifies most of turns containing COBs as "Provide Explanation/Justification" and "Suggest". This does not, of course, prove that there is a statistical correlation between these argumentative categories and COBs because the dataset does not contain negative examples, i.e. turns that are not

COB-annotated are not necessarily those that not contain COBs. We have constructed a mapping table, shown in Table 4, which maps argumentative categories onto a numerical scale of cooperativeness. With this mapping, a large proportion of turns that received COB annotation receive positive cooperativeness score.

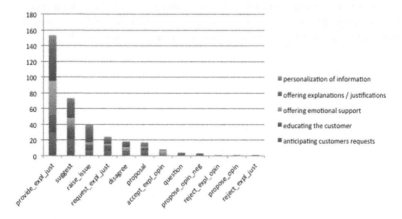

Fig. 3 Correlation between argumentative categories and customer-oriented behaviors.

Table 4 Mapping table for argumentative categories to levels of cooperativeness

Argumentative Categories	Level of Cooperativeness
Accept explanation	5
Suggest	4
Propose	3
Provide opinion	2
Provide explanation or justification	1
Request explanation or justification	0
Question	-1
Raise issue	-2
Provide negative opinion	-3
Disagree	-4
Reject explanation or justification	-5

The cooperativeness score is a measure obtained by averaging the score obtained by mapping argumentative labels of each turn in the conversation into a [-5 +5] scale. The mapping is hand crafted and inspired by Bales's Interaction

Process Analysis framework (Bales, 1950), where uncooperativeness (i.e. negative scores) is linked to high level of criticism, which is not balanced by constructive contributions (e.g. suggestions and explanations). This mapping provides a reasonable indicator of controversial (i.e. uncooperative) conversations.

In Fig. 4, we can observe how COBs are distributed over cooperativeness score. This provides us with an intuition that cooperativeness score can be a predictor of COBs. We need to stress however that from the dataset we cannot assume that those turns that are not COB annotated do not actually contain COBs.

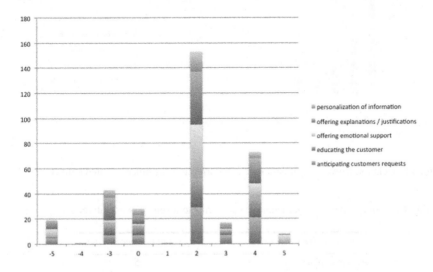

Fig. 4 Distribution of COB annotated turns over cooperativeness scores.

Since the corpus were only partially COB annotated, the only conclusion we can draw from this dataset is that when a turn is COB annotated, there is a high chance that the A3 algorithm would classify it with an argumentative category, which is in turn mapped into a positive cooperativeness score. We cannot conclude instead that turns receiving argumentative annotations that are mapped onto negative cooperativeness scores would actually correspond to absence of COBs.

Our, so far not validated but highly intuitive assumption is that negative cooperativeness scores are also predictors of absence of COBs. This means that turns receiving negative cooperativeness scores can be recognized as elements of customer dissatisfaction. We have reviewed the dataset for those turns that received negative cooperativeness scores and realized that this intuition was reasonably justified. We will provide a more formal assessment of this hypothesis in future work once we have annotated the remaining portion of the dataset.

4.3 Interaction Analytics for Call Center Conversations

We used the Tableau[8] visualization system, which revealed to be a suitable tool for getting insightful multi-dimensional aggregations. We assembled the charts into dashboards that can be used to obtain appropriate summarized information to address the four objectives mentioned in the beginning of this section. We will review each of these objectives and present the related generated dashboard from the analyzed call center data.

Identify Customer Satisfaction

The implementation of the Customer Satisfaction KPI (CSAT) is a direct consequence of the ability to predict COBs. In fact, from Rafaeli et al. (2007), a positive significant correlation exists between customer ratings and COBs.

We pushed this concept a little bit further and we crafted a combined CSAT score by combining cooperativeness score, sentiment and subjectivity analysis. Then turn scores are averaged over the whole conversation. The idea is that customer satisfaction is the result of the overall interaction between the customer and the representative, and not just the occurrence of certain words with positive or negative connotation. We believe that our implementation of CSAT will prove to be more effective and accurate than those based on Sentiment Analysis only.

Combined with additional extracted information such as Sentiment and Subjectivity (see Pallotta and Delmonte (2011) for details), we might safely conclude that CSAT can be predicted by argumentative analysis. Unfortunately, we cannot provide a fully quantitative proof for this claim as the data were not fully annotated and negative examples (i.e. customer dissatisfaction) are missing. As we already said, we consider refining our study towards this direction in future work.

Identify Root Cause of Problems

By looking at controversial topics we can identify root cause of problems in call centers. We selected the worst 20 topics ranked according to frequency of negative attitudes obtained by the Sentiment Analysis module. Fig. 5 shows a dashboard that can be used to detect controversial topics and thus help in spotting unsolved issues.

The user can select one topic and display who addressed that topic (agent or customer) and see the cooperativeness score of each speaker. The rationale in crossing sentiment information with cooperativeness score is a better understanding of the context of sentiment analysis. If, for instance a negative word is used in cooperative context (e.g. "providing an explanation") then its impact in the determination of a cause of problem should be diminished.

---

[8] http://www.tableausoftware.com

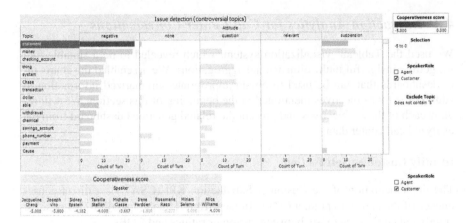

Fig. 5 Problem spotting dashboard

For instance, if the representative says: "in order to avoid this problem, you should remove the virus from the computer", clearly this cannot be considered a negative statement because the representative is simply providing an explanation. This example highlights the limitations of current sentiment-based Speech Analytics solutions, which might be overcome by adopting our approach.

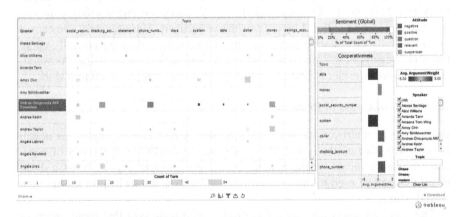

Fig. 6 Topic and Behaviors Dashboard

Another useful tool is that shown in Fig. 6. This dashboard highlights the top 10 most discussed topics and how cooperatively speakers discuss these topics. In the main pane, rows correspond to speakers and for each topic the level of cooperativeness is displayed as a square whose dimension represents the number of turns centered around that topic and the color represents the cooperativeness score. The cooperativeness histogram shows the overall cooperativeness score for each topic and it is refined when the user selects a specific speaker.

Identify Problematic Customers

A critical issue in this domain is that customers are not all the same and need to be treated differently according to their style of interaction. There are agents with interpersonal skills who are able to comfortably deal with demanding customers. Agents who show consistently positive cooperativeness can be assumed to be more suitable to deal with extreme cases. Customers who have already shown negative or uncooperative attitudes could be routed to more skilled agents in order to maximize the overall call center performance (i.e. customer satisfaction).

We present a dashboard where problematic customers can be identified and given a particular care. The main tool for this task is a dashboard for speaker assessment shown in Fig. 7. With this dashboard speakers (agents or customers) are ranked according to their cooperativeness score. In the right-hand pane, also the sentiment analysis results are displayed and compared to the overall sentiment score. The analyst can then drill through a specific customer and visualize a specific customer and the calls he/she made.

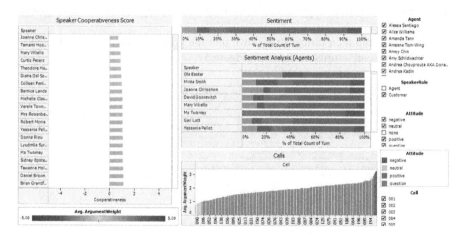

Fig. 7 Speaker Assessment Dashboard

Once drilled down to a specific call, another useful tool for enabling the detection of problematic customers is the conversation graph (Ailomaa et al. 2008), also integrated in our system. The dashboard shown in Fig. 8 reveals interesting facts about the selected call. In the lower pane, calls are ranked according to their average cooperativeness score. By looking at specific calls, the analyst can display the conversation graph that plots the interaction over the call's timeline (i.e. turns on X-axis). The Y-axis represents the cooperativeness score of each turn. In the right pane, Sentiment and Argumentative breakdowns are presented for the selected call. Looking closely at calls with conversation graphs helps the supervisor to understand some interaction patterns. If the customer asks for explanations and the representative fails in providing them, the call's cooperativeness score will be lower. The intuition behind this is that challenging turns must be balanced by

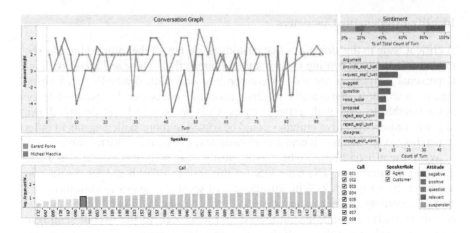

Fig. 8 Conversation graph dashboard

collaborative turns (e.g. explanations must be provided, suggestions must be given to raised issues).

The analyst can then further drill down into the graph or other charts' elements and look at the call's turns as shown in Figure 5.

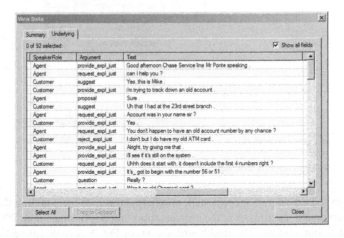

Fig. 9 Drilling through the call's turns

With a different type of representation, our analysis can also serve for exception-based quality management. With the help of a *control chart* displayed in Fig. 10, we can have an overall look of calls quality (measured by the cooperativeness score) and be warned of the presence of outliers (i.e. calls whose cooperativeness score falls outside the control limits [-3σ, +3σ], which can be in turn analyzed in more detail.

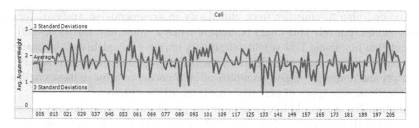

Fig. 10 Control chart of cooperativeness score

Learn Best Practices from Conversations

The implementation of this objective results from considering all the insights gained through the presented visualizations. In particular, Fig. 7 with Agent filtering activated allows one to visualize overall and specific agent's behavior. Best scoring agents can be taken as models and their interaction used as models.

While most of available solutions for skill-based inbound call routing are based on ACD information such as area codes for agent's language selection or based on IVR[9] for option selection. Additionally, the agent selection is often based on efficiency measures in order to optimize the costs and workload (e.g. by assigning the fastest agent to the longest queue). If this strategy might maximize efficiency, they are insufficient to maximize customer satisfaction. We advocate for skill-based call routing based on interpersonal qualities and by influencing the agent selection by cooperativeness requirements.

5 Conclusions

In this article we have presented a new approach to Customer Interaction Analytics based on Interaction Mining, contrasting the current approach based on Speech Analytics and Text Mining. We presented an Interaction Mining tool, which is built on pragmatic analysis of conversations based on argumentation theory. This tool allows us to automatically annotate turns in transcribed conversations with argumentative categories by highlighting the argumentative function of each turn in the conversation. We also showed that our system is robust enough to deal with automatically transcribed speech, as it would be the case in the Call Center domain. An experiment was conducted to see the impact of this technology to a real case. We applied our tool to a corpus of transcribed call center conversations in the banking domains and presented the extracted information in several fashions with the goal of implementing relevant KPIs for Call Center Quality Management.

As for future work we would like to explore other pragmatic dimensions beyond argumentation. This might be relevant in the Call Center domain to look at COBs that are more related to emotional support or providing personalized information, which do not directly relate to argumentation.

---

[9] Interactive Voice Response.

We need also to consider finer granularity in argumentative analysis, for instance at clause level. This might be helpful when a single turn carries several argumentative functions. This would definitively improve the quality of the analysis. Our goal is to implement other KPIs for the Call Center domain such as adherence to scripts and corporate image exemplification. In order to achieve these challenging objectives, new types of pragmatic analysis will be required.

Additionally, we would like to explore the possibility of automatically learning agent and customer profiles from our analysis in order to implement more effective skill-based call routing.

Finally, we will annotate the corpus with negative examples, namely turns that do not contain Customer Oriented Behavior. However, the simple absence of COBs does not automatically entail that the turn contains behaviors that could be perceived by the customers as uncooperative. Therefore, a new model will be required where a new class of behavior needs to be identified to model behavior that might lead to customer dissatisfaction. Unfortunately, the corpus does not contain conversations that received low customer ratings. Probably even more challenging will be the task of finding a corpus of call center conversations that contains them.

References

Ailomaa, M.: Answering Questions About Archived, Annotated meetings. PhD thesis N° 4512, Ecole Polytechnique Fédérale de Lausanne (EPFL), Switzerland (2009)

Baird, H.: Ensuring Data Validity Maintaining Service Quality in the Contact Center. Telecom Directions LLC (2004)

Delmonte, R.: Computational Linguistic Text Processing – Logical Form, Semantic Interpretation, Discourse Relations and Question Answering. Nova Science Publishers, New York (2007)

Delmonte, R.: Computational Linguistic Text Processing – Lexicon, Grammar, Parsing and Anaphora Resolution. Nova Science Publishers, New York (2009)

Delmonte, R., Bristot, A., Pallotta, V.: Deep Linguistic Processing with GETARUNS for spoken dialogue understanding. In: Proceedings of LREC 2010 Conference, Malta, pp. 18–23 (May 2010)

Delmonte, R., Pallotta, V.: Opinion Mining and Sentiment Analysis Need Text Understanding. In: Pallotta, V., Soro, A., Vargiu, E. (eds.) DART 2011. SCI, vol. 361, pp. 81–95. Springer, Heidelberg (2011)

Dunning, T.: Accurate methods for the statistics of surprise and coincidence. Computational Linguistics 19(1), 61–74 (1993)

Feldman, R., Sanger, J.: The Text Mining handbook. Advanced approaches in analyzing unstructured data. Cambridge University Press (2006)

Fiscus, J.G., Ajot, J., Garofolo, J.S.: The Rich Transcription 2007 Meeting Recognition Evaluation. In: Stiefelhagen, R., Bowers, R., Fiscus, J.G. (eds.) RT 2007 and CLEAR 2007. LNCS, vol. 4625, pp. 373–389. Springer, Heidelberg (2008)

Hakkani-Tür, D.: Towards automatic argument diagramming of multiparty meetings. In: Proceedings of the IEEE International Conference on Acoustics, Speech, and Signal Processing (ICASSP), Taipei, Taiwan (April 2009)

Gavalda, M., Schlueter, J.: The Truth is Out There: Using Advanced Speech Analytics to Learn Why Customers Call Help-line Desks and How Effectively They Are Being Served by the Call Center Agent. In: Neustein, A. (ed.) Advances in Speech Recognition: Mobile Environments, Call Centers and Clinics, ch. 10. Springer Media, LLC (2010)

Gilman, A., Narayanan, B., Paul, S.: Mining call center dialog data. In: Zanasi, A., Ebecken, N.F.F., Brebbia, C.A. (eds.) Data Mining V. WIT Press (2004)

Hain, T., Burget, L., Dines, J., Garner, P.N., El Hannani, A., Huijbregts, M., Karafiat, M., Lincoln, M., Wan, V.: The AMIDA 2009 Meeting Transcription System. In: Proceedings of INTERSPEECH 2009, Makuhari, Japan, pp. 358–361 (2009)

Janin, A., Baron, D., Edwards, J., Ellis, D., Gelbart, D., Morgan, N., Peskin, B., Pfau, T., Shriberg, E., Stolcke, A., Wooters, C.: The ICSI Meeting Corpus. In: The ICSI Meeting Corpus. In: Proceedings of IEEE/ICASSP 2003, Hong Kong, April 6-10, vol. 1, pp. 364–367 (2003)

Minnucci, J.: Call Center KPIs: A Look at How Companies Are Measuring Performance, a special report published by ICMI Inc., p. 2 (2004)

Neustein, A. (ed.): Advances in Speech Recognition: Mobile Environments, Call Centers and Clinics. Springer Media, LLC (2010), doi:10.1007/978-1-4419-5951-5_10

Pallotta, V.: Framing Arguments. In: Proceedings of the International Conference on Argumentation ISSA, Amsterdam, Netherlands (June 2006)

Pallotta, V., Delmonte, R.: Argumentative Models for Interaction Mining. Journal of Argument and Computation 2(2) (2011)

Pallotta, V., Seretan, V., Ailomaa, M.: User requirements analysis for Meeting Information Retrieval based on query elicitation. In: Proceedings of the 45th Annual Meeting of the Association for Computational Linguistics, ACL 2007, Prague, pp. 1008–1015 (2007)

Pallotta, V., Vrieling, L., Delmonte, R.: Interaction Mining: Making business sense of customers conversations through semantic and pragmatic analysis. In: Zorrilla, M., Mazón, J.-N., Ferrández, Ó., Garrigós, I., Daniel, F., Trujillo, J. (eds.) Business Intelligence Applications and the Web: Models, Systems and Technologies. IGI Global (2011) (to appear)

Rafaeli, A., Ziklik, L., Doucet, L.: The impact of call center employees' customer orientation behaviors on customer satisfaction. Journal of Service Research (2007)

Rienks, R., Verbree, D.: About the usefulness and learnability of argument-diagrams from real discussions. In: Proceedings of the 3rd International Machine Learning for Multimodal Interaction Workshop (MLMI 2006), May 1-4, Bethesda, MD, USA (2006)

Seretan, V., Wehrli, E.: Multilingual collocation extraction with a syntactic parser. Language Resources and Evaluation 43(1), 71–85 (2009)

Tang, M., Pellom, B., Hacioglu, K.: Call-type classification and unsupervised training for the call center domain. In: Proceedings of the IEEE Workshop on Automatic Speech Recognition and Understanding ASRU 2003, pp. 204–208 (2003)

Takeuchi, H., Subramaniam, L.V., Nasukawa, T., Roy, S.: Getting insights from the voices of customers: Conversation mining at a contact center. In: Information Sciences, vol. 179, pp. 1584–1591. Elsevier (2009)

Zweig, G., Siohan, O., Saon, G., Ramabhadran, B., Povey, D., Mangu, L., Kingsbury, B.: Automated Quality monitoring for call centers using speech and NLP technologies. In: Proceedings of the Human Language Technology Conference of the NAACL, Companion Volume, pp. 292–295. Association for Computational Linguistics, New York City (2006)

A Linguistic Approach to Opinion Mining

Franco Tuveri and Manuela Angioni

Abstract. Reviews are used every day by common people or by companies who need to make decisions. Such amount of social data can be used to analyze the present and to predict the near future needs or the probable changes. Mining the opinions and the comments is a way to extract knowledge by previous experiences and by the feedback received. In this chapter we propose an automatic linguistic approach to Opinion Mining by means of a semantic analysis of textual resources and based on FreeWordNet, a new developed linguistic resource. FreeWordNet has been defined by the enrichment of the meanings expressed by adjectives and adverbs in WordNet with a set of properties and the polarity orientation. These properties are involved in the steps of distinction and identification of subjective, objective or factual sentences with polarity valence and contribute in a basic way to the task of features contextualization.

Keywords: NLP, WSD, Opinion Mining, Text Categorization.

1 Introduction

Currently millions of people use regularly social networks to communicate, to share information and interests and to express opinions about any topic in form of reviews of product on blogs, forum, and discussion groups. Reviews are used every day by common people or by companies who need to make decisions. They facilitate to book a hotel, to buy a book, or to taste the market tracing the customer satisfaction about a product. Mining the opinions and the comments is a way to extract knowledge by previous experiences and by the feedback received. The opinion monitoring activity is essential for listening to and for taking advantage of the conversations of possible customers. Currently the main objective is to move

Franco Tuveri · Manuela Angioni
CRS4, Center of Advanced Studies,
Research and Development in Sardinia,
Parco Scientifico e Tecnologico, Ed. 1,
09010 Pula (CA), Italy
e-mail: {tuveri,angioni}@crs4.it

C. Lai et al. (Eds.): New Challenges in Distributed Inf. Filtering and Retrieval, SCI 439, pp. 113–129.
springerlink.com   © Springer-Verlag Berlin Heidelberg 2013

forward the frontier of the Opinion Mining passing through the simple evaluation of the polarity of the expressed feeling, to a one where extracted sentiments are related to the context and the information about the features is more detailed.

The term feature is used with the same sense given by Ding et al. (2008) in their approach to Opinion Mining. Given an object, that could be a service, a person, an event or an organization, the term feature is used to represent a component or an attribute describing that object. In the chapter the reviews are related to a specific object that needs to be contextualized in order to more easily combine the qualitative information expressed by the users about the highlighted features.

The chapter describes an Opinion Mining system able to automatically extract the features and the meaningful information contained in opinions, independently by the domain. For example, with regard to politics, it could be very relevant investigate how the electorate reacts to the affirmation of a politician in relation to an event. Certainly in such case features could not be determined a priori, but should be extracted from the text automatically.

As in (Benamara et al., 2007), we propose a linguistic approach to Opinion Mining. In this respect we developed FreeWordNet, a linguistic resource based on WordNet (Miller, 1995), because of the lack in existing resources of relevant information that contributes to the information contextualization. FreeWordNet has been performed by the enrichment of adjectives and adverbs' synsets with a set of properties related both to the emotional/moody sphere, but even to the geographical, behavioral, physical and other related spheres. The analysis of the opinions is performed through the processing of textual resources, the information extraction and the evaluation and the summarization of the semantic orientation of the sentences. More in details, the approach performs a semantic disambiguation and a categorization phase and takes into account the meanings expressed in conversations, considering for instance the synsets and the semantic relations related to relevant terms.

A Semantic Classifier (Angioni et al., 2008) performs the categorization of the corpus of reviews, producing as result the most representative categories and the weights as measures of their relevance in order to define the domain of the features in the review. The resulting categories provide a first level of contextualization and a support for the activities of the Word Sense Disambiguation (WSD). A semantic similarity algorithm performs the WSD based on a modified version of the measure proposed by Leacock-Chodorow (Leacock and Chodorow, 1998) and on the mapping of the synsets on WordNet Domains categories (Magnini et al., 2002) (Magnini and Strapparava, 2004). The algorithm produces as result a matrix of weights that defines a map of the existing relations between features and groups them in more specific thematic set. The syntactic and semantic analyses are performed in order to achieve a distinction between objective and subjective phrases. Factual sentences having polarity are also identified in order to distinguish between opinions and facts having positive or negative value. The identification of adjectives and adverbs, the extraction of information by means of their associated properties and FreeWordNet has a relevant role in this phase.

The chapter will offer more details about the reasons behind the development and the method used to build FreeWordNet showing some evaluation measures and a comparison with other similar available resources. Finally more details about the feature extraction and the opinion summarization will be produced.

The remainder of the chapter is organized as follows: Section 2 refers to related works. Section 3 introduces the Opinion Mining approach and examines the work performed on the preparation of data, giving some details about the semantic classification and the contextualization of features. Moreover, it explains the development of FreeWordNet, showing some measures and comparisons. Section 4 describes in details the feature extraction process and the approach to the opinion extraction by means of the identification of associations between adjectives, adverbs and features. Section 5 illustrates an informal test case and, finally, Section 6 draws conclusions.

2 Related Works

Many approaches to Opinion Mining and Sentiment Analysis are based on linguistic resources, lexicons or lists of words, used to express sentiments or opinions. The lack of suitable and/or available resources is one of the main problems in the Opinion Mining process and in general in the analysis of textual resources by the application of Natural Language Processing techniques. Knowing the polarity of words and their disambiguated meanings can surely help to better identify the opinions related to specific features. In (Wiebe and Mihalcea, 2006) the authors evidenced that subjectivity is a property to be associated to word senses and that WSD can "directly benefit from subjectivity annotations".

SentiWordNet (Esuli and Sebastiani, 2007) (Baccianella et al., 2010) is one of the publicly available lexical resources that extends WordNet thanks to a semi-automatic acquisition of the polarity of WordNet terms, evaluating each synset according to positive, negative and objective values. It provides the possibility to accept user feedback on the values assigned to synsets, allowing the building of a community of users in order to improve SentiWordNet. Despite its wide coverage, SentiWordNet does not give information useful for the contextualization of the opinions.

A further resource is MicroWnOp (Cerini et al., 2007). Used in order to validate SentiWordNet, MicroWnOp is a carefully balanced set of 1,105 WordNet synsets manually annotated by a group of five human annotators according to their degrees of polarity with the three scores summing up to 1. Synsets 1-110 have been tagged by all the annotators working together. A group of three of the five annotators tagged the synsets 111-606, and the remaining two annotators tagged all synsets 607-1,105.

MicroWnOp has been built as a "Gold Standard" adopting two criteria have been adopted:

- Opinion relevance: the synsets should be relevant to represent the opinion topic.
- WordNet representativeness, respecting the distribution of synset among the four parts of speech.

Q-WordNet (Agerri and García-Serrano, 2010) is another lexical resource, consisting of WordNet senses automatically annotated by positive and negative polarity. Q-WordNet tries to maximize the linguistic information contained in WordNet, taking advantage of the human effort given by lexicographers and annotators.

WordNet-Affect (Valitutti et al., 2004), available for free for non-profit institution, has been developed starting from WordNet, assigning one or more affective labels (a-labels) to a subset of synsets representing affective concepts that contribute to precise the affective meaning. For example, the a-label Emotion represents the affective concepts related to emotional state. Other concepts are not emotional-affective but represent moods, situations eliciting emotions, or emotional responses. The same staff developed WordNet Domains, a resource that maps the WordNet synsets to a subset of categories of the Dewey Decimal Classification System.

In opinion summarization several approaches are based on the use of lexicons of words able to express subjectivity, without considering the specific meaning the word assumes in the text by means of any form of semantic disambiguation.

Other approaches consider instead the word meanings as (Akkaya et al., 2009), that build and evaluate a supervised system to disambiguate members of a subjectivity lexicon, or (Rentoumi et al., 2009), that propose a methodology for assign a polarity to word senses applying a WSD process.

Some authors (Lee et al., 2008) asserted that the introduction of the sense disambiguation in text analysis showed that systems adopting syntactic analysis techniques on extracting opinion expressions tend to show higher precision and lower recall than those which do not adopt this kind of techniques.

Feature extraction is a relevant task of the opinion summarization process. Some works about features are based on the identification of nouns through the pos-tagging and provide an evaluation of the frequency of words in the review based on tf-idf criterions and its variation (Scaffidi, 2007). (Hu and Liu, 2004) proposed a very promising study about opinion summarization. The objective of the study, based on data mining and natural language processing methods, is to provide a feature-based summary of a large number of customer reviews of some products sold online.

Others researchers (Zhai et al., 2010) proposed a constrained semi-supervised learning method based on the contextualization of reviews grouped in specific domains. The method also try to solve the problem to group feature expressions and to associate them to feature labels using a characterization of the features defined by users. They do not use WordNet for several reasons including the problem of the semantic disambiguation, the lack of technical terms or specific meanings related to the context of use, or yet the differences of synonymy between different context.

Finally another important work is (Popescu and Etzioni, 2005), that worked on the explicit features in noun phrases.

3 The Opinion Mining Approach

As asserted by Lee et al. (2008), "Opinion Mining can be roughly divided into three major tasks of development of linguistic resources, sentiment classification, and opinion extraction and summarization".

Our linguistic approach to Opinion Mining embraces the above division into the main tasks and focuses on the feature extraction and on the role of FreeWordNet.

3.1 Preparation of Data

Some research has been conducted about opinion retrieval systems in order to collect reviews and opinions from blogs about a subject specified in queries. An opinion retrieval system may be implemented by enhancing an information retrieval system with opinion finding algorithms based on a lexicon of subjective words. In (Zhang et al., 2007) has been proposed an algorithm that retrieves blog documents according to the opinions.

In this chapter we evidence the phase of the preparation of a corpus of data defined as a set of results about a specific topic. This step provides in output a set of information about the corpus, the set of reviews, the disambiguated terms related to each sentence and review, and the relations between adjectives, adverbs and terms and we do not mind about the way reviews are gathered from information sources. Considering a specific domain a set of reviews are collected and organized as a file. In the use case, described in Section 5, the tourism domain has been chosen, in order to evaluate the system.

The first step has been to collect reviews discarding or correcting sentences having orthographic errors. Only well-built sentences have been selected and inserted in the corpus in order to avoid introducing errors and to facilitate the syntactic parser activities.

The reviews have been managed independently in order to maintain the relations between features, comments and opinions expressed by the users.

The preparation of data involves some linguistic resources:

- WordNet, the linguistic resource developed by the Princeton University
- WordNet Domains, the lexical resource representing domain associations among the word-senses of WordNet and a subset of categories of the Dewey Decimal Classification (DDC) system
- FreeWordnet, an extension of the properties of a subset of adjectives and adverbs contained in WordNet

and some tools for the syntactic analysis and the semantic categorization of the opinions.

TreeTagger (Schmid, 1994) has been used in order to perform the syntactic analysis of the sentences and the phrase chunking process, annotating text with part-of-speech tags and lemma information and identifying into each sentence its sub-constituents such as noun chunks. A Java class wraps the evaluation provided by the TreeTagger parser and chunker and, analyzing the parts of speech, identifies the associations between nouns and their related information.

A Semantic Classifier, involved in several tasks, also plays a relevant role. It is capable to automatically categorize text documents using the same set of categories of WordNet Domains. The Semantic Classifier categorizes the set of reviews in order to define the related topic and providing as result a set of categories and weights. The categories define the domain for the corpus of reviews. The Semantic Classifier also parses and categorizes each sentence in order to decide which specific phrases are relevant to the topic. The Classifier individually manages each sentence in order to give more specific information in the WSD. For example, analyzing reviews about tourism and especially reviews about hotels, we expect to examine sentences containing opinions about geographical locations, buildings, rooms, staff and food. The semantic categorization task improves the WSD of adjectives and adverbs and the concordance of their meaning with the related feature terms. The gloss associated to each synset is categorized with the aim to relate feature terms and adjectives and/or adverbs. A function defines the semantic distance between the features and the most probable meaning of related adjectives and/or adverbs. The same function is used in the WSD of feature terms.

FreeWordNet is involved in the steps of distinction and identification of subjective, objective or factual sentences and in the analysis and validation of the features extracted in the contextualization phase, useful in referring adjectives and adverbs to the features.

The sentiment classification, in general, concentrates on establishing the orientation of opinions expressed about a particular topic. In our approach, the sentiment classification is carried out in several steps passing through: the linguistic analysis of reviews, the semantic categorization and finally the identification of both subjective and factual sentences having polarity valence.

3.2 The Linguistic Resource

Sentences containing opinions are defined as subjective according to Akkaya et al. (2009) definition: "words and phrases used to express mental and emotional states, such as speculations, evaluations, sentiments, and beliefs". This definition has been extended in FreeWordNet by adding further information having polarity orientation. An example is the information related to the human senses, like the sense of touch and the sense of taste.

FreeWordNet, also, represents an answer to well defined needs arising from the development of an automatic process of feature extraction. It is involved in the steps of distinction and identification of subjective, objective or factual sentences and contributes in a basic way to the task of features contextualization.

Some considerations and evaluation about existing linguistic resources based on WordNet, have been made before the development of FreeWordNet.

We found that some additional properties were necessary in the analysis of sentences in order to identify relations between nouns and adjectives and to improve the analysis of the performances of our tools in the WSD phase.

In FreeWordNet, that will be soon freely available, adjectives and adverbs have two different levels of categories. The first level is automatically performed by the semantic categorization of the glosses of the terms and assigns to each synset a subset of categories provided by WordNet Domains, useful in order to associate features to adjectives and adverbs. The second level, related to the human categorization, defines sets of 14 and 7 categories, respectively for adjectives (Table 1) and adverbs (Table 2). We suggest these second sets of categories although we are aware of the fact that some areas are still not covered, e.g. the economic sphere. The idea was that adjectives and adverbs could be grouped in categories according to their meanings. Together the two levels of categories define the context of use of the terms in a sentence. For example, in the Tourism domain, if an adjective has Moral/Ethic property and is categorized as Person, we can expect the context to be related to personnel evaluation.

This kind of categorization is used in order to distinguish between subjective and factual sentences through a polarity evaluation. Categories that identify properties related to the moral/ethic or emotional sphere suggest subjective values, while others, that identify e.g. chronologic or shape categories imply factual valence.

Table 1 Adjectives' properties

Adjectives	Pos	Neg	Obj	Tot
Emotion	52	73	3	128
Moral/Ethic	45	155	2	202
Character	355	584	220	1159
Weather	7	26	6	39
Color	0	9	42	51
Quantity	16	0	9	25
Appearance	41	83	46	170
Material	22	11	54	87
Shape	0	0	30	30
Touch	3	13	6	22
Taste	40	41	5	86
Dimension	11	2	60	73
Chronologic	3	0	30	33
Geographic	0	10	19	29
Others	29	17	87	133
Tot. Adjectives	624	1024	619	2267

We agree with Agerri and Serrano (2010) considering that the polarity classification involves not only subjective but even factual sentences in the detection of polarity values of word senses in texts. A very relevant note is that polarity sentences can be distinguished in sentences having subjective values and factual sentences having polarity values. The first group is composed by phrases expressing opinions while the other describes situations, facts and might involve word senses or sentences objectively having polarity valence.

An example of factual sentence could be the following: "The season is arid". This is not a subjective but a factual sentence and expresses a negative meaning from the point of view of the life. The adjective "arid", having synset 02552415 in WordNet 3.0, has meaning specified by this gloss: "lacking sufficient water or rainfall". The comparison between different resources evidences some differences in the polarity evaluation. In SentiWordNet the synset is evaluated as negative with score 0.375 and objective with score 0.625 and not gives any information about factual information, while in FreeWordNet the same synset has a negative polarity with the property "Weather" associated.

The human categorization of adjectives and adverbs provides the possibility to separate sentences having polarity orientation from the others and to distinguish between adjectives having subjective meaning and adjectives having factual valence. Categories like Emotion, Moral/Ethic, Character, Taste, Touch, Appearance, have a subjective orientation. The categories Weather, Color, Quantity, Material, Dimension, Chronologic, and Geographic express factual polarity values. Table 1 depicts the list of the properties related to adjectives and the number of synsets having positive, negative or objective valence. Factual sentences having polarity give information about situations or facts that could be evaluated as positive or negative according to objective criteria related to their description, while subjective sentences give personal opinions about features, facts, events, etc.

Table 2 Adverbs' properties

Adverbs	Pos	Neg	Obj	Tot
Time	0	0	7	7
Manner(things)	18	25	8	51
Manner(persons)	166	205	5	376
Place	0	0	3	3
Intensifiers	0	0	38	38
Quantity	0	0	6	6
AND	1	0	0	1
Total Adverbs	185	230	67	482

Regarding the adverbs, their meaning, position and strength have been taken in consideration. Their categorization has been made distinguishing between adverbs of manner, adverbs of place, adverbs of time, adverbs of quantity or degree, of affirmation, negation or doubt (grouped as AND adverbs), adverbs as intensifiers or emphasizers and adverbs used in adversative and in consecutives sentences, as listed in Table 2.

Only the adverbs of manner may be positive or negative. The adverbs of degree give the idea about the intensity with which something happens or have an impact on the expressed opinions. The other adverbs give additional information to the analysis related to the location, the direction or the time.

Some studies (Pang et al., 2002) demonstrated that, as adjectives and adverbs, other parts of speech contribute to express opinions or sentiments. In their study on movie-review polarity classification the researchers affirm that nouns and verbs can also be strong indicators for sentiment. The current version of FreeWordNet includes only adjectives and adverbs.

The comparison of the measures with similar resources like SentiWordNet, Q-WordNet and MicroWNOp evidenced some limits in FreeWordNet. The most relevant is given by the limited number of synsets included in the resource. For this reason, the evaluation of results in the test phase evidenced that some adjectives and adverbs has not been detected. As result, some sentences have not been correctly analyzed because the text analysis is strongly affected by the recognition of adjectives and adverbs having polarity valence.

Moreover, the results point out that several categories of adjectives need to be enriched. Regarding the adverbs, the test phase puts in evidence the need for an adequate coverage for the chosen corpus and domain.

We would like to stress that FreeWordNet is still under development. Despite this, the additional properties associated to the synsets can bring relevant benefits to the analysis of opinions. For example the distinction between subjective and factual adjectives and adverbs having polarity values is an implicit capability of FreeWordNet.

3.3 Opinion Extraction and Summarization

The term "opinion-oriented information extraction" (opinion-oriented IE) or simply "opinion extraction" refers to information extraction problems specific for Sentiment Analysis and Opinion Mining (Pang and Lee, 2008). The opinion extraction shouldn't be interpreted in a narrow sense as focused only on the extraction of opinion expressions. For instance, other tasks, as the feature extraction and the polarity evaluation of the opinion expressed, are parts of this task. Opinion Mining has often been used to describe the context of the analysis of reviews related to items, such as products, services, or events. Extracting and analyzing opinions associated with each individual aspect are ultimate tasks in the informative summarization.

The feature extraction, as task of the opinion extraction, is the identification of feature terms and could be considered a standard problem of entity recognition related to the recognition of features having opinions related. The problem of opinion extraction could also be considered as a categorization problem for small portions of text (the reviews) in order to determine whether or not a text is classified as subjective or factual with a polarity orientation associated. All these characteristics describe the opinion extraction activities and include the tasks of text categorization, entity recognition, semantic analysis and analysis of opinions in order to identify features and relations with opinions.

4 Our Approach to Feature Extraction

The feature extraction process is briefly described by the following steps: identification of the context for the domain of the corpus; distinction between subjective and objective or factual sentences; identification of features by means of the semantic analysis and evaluation and validation of the features extracted; creation of a matrix of relations, by means of the assignment of a weight to each possible pair of features, in order to define thematic sets of features; reference of adjectives and adverbs to feature terms.

The identification of a context in Opinion Mining is not only a necessary task in the identification of features but even in the interpretation of the expressed opinions.

Engström (2004) affirms that sentiment in different domains can be expressed in very different ways and that the domain of the items to which it is applied can influence the accuracy of sentiment classification. A reason is that the same phrase can indicate different sentiment in different domains acquiring positive or negative valence in different contexts. For example we can consider the Turney's (2002) observation that the "unpredictable" adjective may have a negative orientation for a car's steering abilities but have a positive meaning in a movie review. This is an issue that brings several authors to evidence the importance of building domain-specific classifiers. The linguistic approach proposed is not domain dependent and can be used with similar performances in every context. For this purpose, the linguistic resource and the categorization task have been introduced as relevant elements involved in the feature extraction process.

The semantic categorization task is a relevant part of the entire process of contextualization. By means of the categorization of the set of reviews, the Semantic Classifier defines a set of domain categories and performs the Word Sense Disambiguation of the features terms and of the related adjectives and adverbs according to the domain categories.

Concerning the categorization task, a syntactic parser and the semantic analysis perform the identification of subjective and objective sentences and the detection of factual sentences having polarity value by means of the identification of adjectives and adverbs. The linguistic resource has a relevant role in this phase. As said, FreeWordNet contains two different levels of categories. The first level is related to the human categorization of adjectives and adverbs and the second level is realized with the categorization of the glosses of WordNet terms, performed by the Semantic Classifier. This second level of categorization, used in WSD and feature contextualization, defines a concordance between features and adjectives and adverbs in opinion expressions. Together the two levels of categories define the context of use of the terms in a sentence.

In this phase, the sentences and the part-of-speech are identified and tagged. The information extracted by the syntactic parser is interpreted in order to put in evidence the relations between the different part-of-speech. The definition of the context contributes to the identification and assignment of the correct sense to the feature terms.

An example of result of the semantic categorization is given by the processing of the sentence: "The arid climate is characterized by a high evaporation and lack of rainfalls". The Semantic Classifier categorizes the sentence and identifies the most relevant categories as showed in Figure 1.

Each sentence of the reviews is categorized in order to distinguish between subjective and objective, with or without orientation, sentences. The distinction aims to define a view of the opinions expressed by users and a collection of objective data about the topic.

The identification of the features is performed with a tf-idf function, calculated on the sentences having polarity orientation that extracts a first list of candidate features. The features are then reduced to a more restricted set by the application of a WSD algorithm, based on a semantic distance function, able to identify the correct meanings of feature terms according to the context.

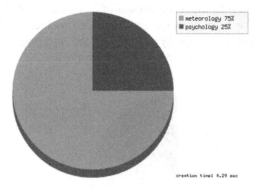

meteorology 75%
psychology 25%

creation time: 0.29 sec

Fig. 1 The result of the semantic categorization

The semantic distance function is a modified version of the Leacock and Chodorow algorithm, and calculates the semantic distance between the synsets related to the features. The distance between synsets is based on the semantic net of WordNet and also depends on the number of synsets related to each feature term and on the common categories of the synsets. More in details, the function assign a weight to each possible pair of features, according to the mapping of their synsets to the set of the topic categories, extracted by the Semantic Classifier. The measure is based on the principle that the bigger is the number of categories in commons, the smaller is the distance between synsets. This implies that two features have a stronger relation the bigger is the weight associated to their relation.

The WSD algorithm, using the values calculated by the semantic distance function, creates a matrix of relations between features. The evaluation based on suitable thresholds' values on the calculated weights of the relations in the mentioned matrix provides selected lists of related features. The rows and columns of said matrix are the extracted features. The matrix contains as weights the distance values that measure the strength of the relations existing between two features. As said, the higher the weight, the stronger the relation.

The adjectives having polarity and the adverbs of manner and the intensifier ones have to be put in relation with the features terms they are referred to. The TreeTagger chunker plays a very relevant role in this phase, as described in the next section.

The property and the polarity value of each adjective with multiple meanings, once disambiguated, enrich the information about the related feature. The presence of intensifier adverbs contributes to determine the grade of the expressed opinion.

The WSD of adjectives, as described in the following example, is extremely important. The adjective "arid" in the sentence: "The arid climate is characterized by a high evaporation and lack of rainfalls" has two meanings, each related to a different gloss categorized with different sets of categories. Both the glosses of the adjective "arid" have been analyzed and classified in FreeWordNet. The right sense of the adjective is chosen by considering the matching of the categories, showed in Figure 1, with the categories of both the glosses.

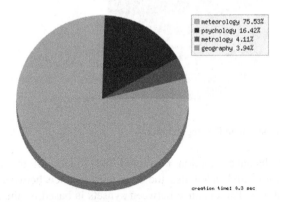

Fig. 2 Semantic categorization of arid

Figure 2 shows the categories related to the synset "302462790" having gloss "lacking sufficient water or rainfall; an arid climate". The main category associated to the synset, "meteorology", matches the same main category of the sentence, providing in such way information about the more probable meaning of the adjective. Its property and value are used in the polarity evaluation of the sentence.

4.1 Referring Adjectives and Adverbs to Features

In order to perform a summarization of the opinions, the association between features and adjectives and adverbs included in the sentences is considered. As said, a wrapper evaluates the results provided by the TreeTagger parser and chunker, analyzing the parts of speech tags and identifying into each sentence its sub-constituents such as noun chunks. As result, it identifies the associations between the features and their related information.

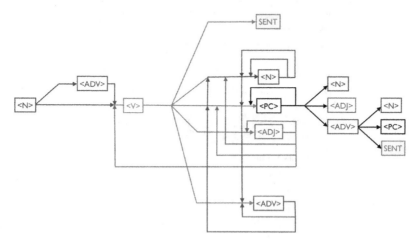

Fig. 3 The possible patterns of chunks

Figure 3 shows the graph implemented in order to identify all possible sequences of chunks in the sentences. The wrapper implements a set of rules, based on the relations depicted in the graph and puts in relation the different chunks. It refers to the parts of speech in order to have a precise association between the features and their related information.

In the Figure 3 N stands for noun, ADJ for Adjective, ADV for Adverb, V for Verb, PC for prepositional chunk, and SENT is the symbol used to indicate the conclusion of the sentence.

The use of the chunker makes easier the production of a feature-based summary of opinions and produces better performances in the definition of the relations between adjective, adverbs and the related features. The semantic distance function here again performs the WSD for the features and for the adjectives and adverbs related. The evaluation of the polarity of the meanings of the adjectives and adverbs has been made in order to give a detailed opinion summarization based on each specific feature.

5 An Informal Test Case

A test case is performed on a collection of about 100 reviews of the Alma Hotel of Alghero (Sardinia, Italy) extracted from TripAdvisor, a travel website of advices and reviews from a large community of travellers. About 950 sentences have been extracted from the above reviews by the tools for the syntactic analysis.

Fig. 4 The feature breakfast and its related data

The first step is the identification of the subjective sentences and of the factual sentences having polarity valence. Then the Semantic Classifier categorizes the sentences and gives back as result a small set of categories, such as Tourism, Person and Gastronomy, with the associated weights, defining the domain of the reviews.

The application of a tf-idf function allows the definition of an initial set of about 450 candidate features, based on the frequency of their use in the collection of reviews.

Then a reduction of the features based on the mapping of the synsets on the WordNet Domains categories has been performed. As final result we obtain a set of about 80 features. As said the semantic distance of synsets and the assignment of a weight to the pairs of features defines the matrix of the relations between the features extracted.

Figure 4 is a snapshot of the demo realized in order to test the feature extraction task and the matrix of relations obtained from the algorithm described in the chapter.

The snapshot shows two definitions for the feature "breakfast" related to the topic tourism: the first extracted from Wikipedia and the second from WordNet.

The matrix of features provides the list of the related disambiguated features. The HTML table in Figure 4 shows the mapping existing between each feature and the related concepts with the associated weights, ordered following the strength of the relations determined by the algorithm. A list of the related images is automatically produced in order to give a better representation of the information about the features.

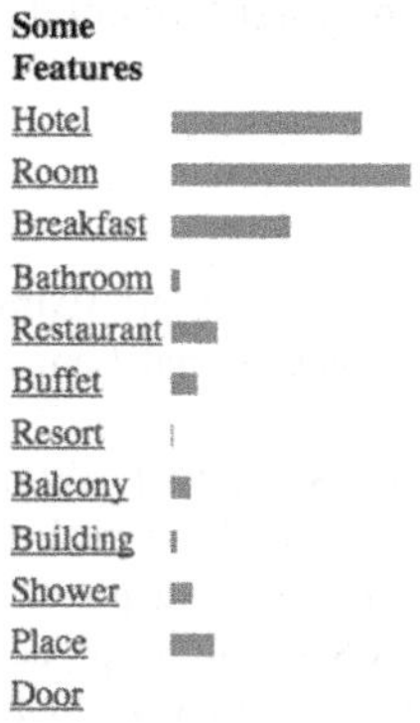

Some Features

Hotel	▰▰▰▰▰▰▰▰▰
Room	▰▰▰▰▰▰▰▰▰▰
Breakfast	▰▰▰▰▰
Bathroom	▮
Restaurant	▰▰
Buffet	▪
Resort	⸽
Balcony	▪
Building	⸽
Shower	▪
Place	▰
Door	

Fig. 5 An example of opinion summarization

Figure 5 shows illustrates a simple polarity evaluation through the evaluation of the preferences expressed about the represented features. In the phase of data aggregation, it is important to notice that a semantic relation between the features breakfast and buffet exists but it does not appear directly in this data representation. It is responsibility of analysts provide the correct interpretation of data. The system currently does not analyze this kind of relations between data, but is able to deduce only that users like the breakfast of the hotel but less the buffet organization.

6 Conclusions and Future Works

Several Opinion Mining methods and techniques have been developed in order to analyze opinions and comments related to services, products, and events. Opinion Mining is essential for listening to and for taking advantage of the conversations of users and it is a way to extract knowledge by the experience and the feedback in order to reach new possible customers.

In this chapter we proposed a linguistic approach to Opinion Mining, relevant in the extraction of feature terms, performing a contextualization of the reviews by means of semantic tools and by the definition of a new linguistic resource. FreeWordNet is involved in the steps of distinction and identification of subjective, objective or factual sentences with polarity valence and contributes in a basic way to the task of features contextualization. Moreover, the use of the synsets and the semantic categorization aim to define a method for the extraction of more accurate meanings and features from textual resources.

Future works will provide the extension of FreeWordNet with new terms and properties, as the measures performed point out that there are several categories of adjectives and adverbs that need to be enriched and the test case denotes the lack of an adequate coverage for the chosen corpus and domain. Moreover, the work will continue with the development of an opinion summarization algorithm, and the generalization of the approach in order to improve its performances. A validation to support the value of the expressed ideas will be one of the goals of the above-mentioned approach and new experimental results will be product.

References

Agerri, R., García-Serrano, A.: 2010. Q-WordNet: Extracting polarity from WordNet senses. In: Seventh Conference on International Language Resources and Evaluation (LREC 2010) (2010)

Akkaya, C., Mihalcea, R., Wiebe, J.: Subjectivity Word Sense Disambiguation. In: Proceedings of the 2009 Conference on Empirical Methods in Natural Language Processing, pp. 190–199. ACL and AFNLP, Singapore (2009)

Angioni, M., Demontis, R., Tuveri, F.: A Semantic Approach for Resource Cataloguing and Query Resolution. In: Communications of SIWN (2008); Special Issue on Distributed Agent-based Retrieval Tools (2010)

Baccianella, S., Esuli, A., Sebastiani, F.: SentiWordNet 3.0: An Enhanced Lexical Resource for Sentiment Analysis and Opinion Mining. In: Proceedings of LREC 2010, 7th Conference on Language Resources and Evaluation, Valletta, MT, pp. 2200–2204 (2010)

Benamara, F., Cesarano, C., Picariello, A., Reforgiato, D., Subrahmanian, V.S.: Sentiment Analysis: Adjectives and Adverbs are better than Adjectives Alone. In: Proceedings of ICWSM 2007 International Conference on Weblogs and Social Media, pp. 203–206 (2007)

Cerini, S., Compagnoni, V., Demontis, A., Formentelli, M., Gandini, C.: Micro-WNOp: A gold standard for the evaluation of automatically compiled lexical resources for opinion mining. In: Sansó, A. (ed.) Language Resources and Linguistic Theory: Typology, Second Language Acquisition, English Linguistics, pp. 200–210. Franco Angeli Editore, Milano (2007)

Ding, X., Liu, B., Yu, P.S.: A Holistic Lexicon-Based Approach to Opinion Mining. In: WSDM 2008 Proceedings of the International Conference on Web Search and Web Data Mining. ACM, New York (2008)

Engström, C.: Topic dependence in sentiment classification. Master's thesis, University of Cambridge (2004)

Esuli, A., Sebastiani, F.: Page Ranking WordNet synsets: An application to Opinion Mining. In: Proceedings of the 45th Annual Meeting of the Association of Computational Linguistics, vol. 45, pp. 424–431. Association for Computational Linguistics (2007)

Hu, M., Liu, B.: Mining and summarizing customer reviews. In: Proceedings of the ACM SIGKDD International Conference on Knowledge Discovery & Data Mining, pp. 168–177. ACM Press (2004)

Leacock, C., Chodorow, M.: Combining local context and WordNet similarity for word sense identification. In: Fellbaum 1998, pp. 265–283 (1998)

Lee, D., Jeong, O.-R., Lee, S.-G.: Opinion Mining of customer feedback data on the web. In: ICUIMC 2008 Proceedings of the 2nd International Conference on Ubiquitous Information Management and Communication (2008)

Magnini, B., Strapparava, C.: User Modelling for News Web Sites with Word Sense Based Techniques. User Modeling and User-Adapted Interaction 14(2), 239–257 (2004)

Magnini, B., Strapparava, C., Pezzulo, G., Gliozzo, A.: The Role of Domain Information in Word Sense Disambiguation. Natural Language Engineering, Special Issue on Word Sense Disambiguation 8(4), 359–373 (2002)

Miller, G.A.: WordNet: A Lexical Database for English. Communications of the ACM 38(11), 39–41 (1995)

Pang, B., Lee, L.: Opinion mining and sentiment analysis. Foundations and Trends in Information Retrieval 2(1-2), 1–135 (2008)

Pang, B., Lee, L., Vaithyanathan, S.: Thumbs up? Sentiment classification using machine learning techniques. In: Proceedings of the Conference on Empirical Methods in Natural Language Processing (EMNLP), pp. 79–86 (2002)

Popescu, A.-M., Etzioni, O.: Extracting Product Features and Opinions from Reviews. In: Proceedings of the 2005 Conference on Empirical Methods in Natural Language Processing, EMNLP 2005 (2005)

Rentoumi, V., Giannakopoulos, G.: Sentiment analysis of figurative language using a word sense disambiguation approach. In: International Conference on Recent Advances in Natural Language Processing (RANLP 2009), Borovets, Bulgaria. The Association for Computational Linguistics (2009)

Scaffidi, C., Bierhoff, K., Chang, E., Felker, M., Ng, H., Jin, C.: Red Opal: product-feature scoring from reviews. In: ACM Conference on Electronic Commerce 2007, pp. 182–191 (2007)

Schmid, H.: Probabilistic Part-of-Speech Tagging Using Decision Trees. In: Proceedings of the International Conference on New Methods in Language Processing, pp. 44–49 (1994)

Turney, P.: Thumbs up or thumbs down? Semantic orientation applied to unsupervised classifi- cation of reviews. In: Proceedings of the Association for Computational Linguistics (ACL), pp. 417–424 (2002)

Valitutti, A., Strapparava, C., Stock, O.: Developing affective lexical re-sources. Psychology 2(1) (2004)

Wiebe, J., Mihalcea, R.: Word Sense and Subjectivity. In: Proceedings of the Annual Meeting of the Association for Computational Linguistics, Sydney, Australia (2006)

Zhai, Z., Liu, B., Xu, H., Jia, P.: Grouping Product Features Using Semi-Supervised Learning with Soft-Constraints. In: Proceedings of the 23rd International Conference on Computational Linguistics (COLING 2010), Beijing, China (2010)

Zhang, W., Yu, C., Meng, W.: Opinion Retrieval from Blogs. In: Proceedings of the ACM Sixteenth Conference on Information and Knowledge Management (CIKM 2007), Lisbon, Portugal (2007)

Sentiment Analysis in the Planet Art:
A Case Study in the Social Semantic Web

Matteo Baldoni, Cristina Baroglio, Viviana Patti, and Claudio Schifanella

Abstract. Affective computing is receiving increasing attention in many sectors, ranging from advertisement to politics. Its application to the Planet Art, however, is quite at its beginning, especially for what concerns the visual arts. This work, set in a Social Semantic Web framework, explores the possibility of extracting rich, emotional semantic information from the tags freely associated to digitalized visual artworks, identifying the prevalent emotions that are captured by the tags. This is done by means of ArsEmotica, an application software that we developed and that combines an ontology of emotional concepts with available computational and sentiment lexicons. Besides having made possible the enrichment of the ontology with over four-hundred Italian terms, ArsEmotica is able to analyse the emotional semantics of a tagged artwork by working at different levels: not only it can compute a semantic value, captured by tags that can be directly associated to emotional concepts, but it can also compute the semantic value of tags that can be ascribed to emotional concepts only indirectly. The results of a user study, aimed at validating the outcomes of ArsEmotica, are reported and commented. They were obtained by involving the users of the same community which tagged the artworks. It is important to observe that the tagging activity was not performed with the aim of later applying some kind of Sentiment Analysis, but in a pure Web 2.0 approach, i.e. as a form of spontaneous annotation produced by the members of the community on one another's artworks.

Keywords: Semantic Web, Ontologies, Sentiment Analysis, Social Tagging.

1 Introduction

Seeing artworks exposed at a vernissage or a gallery has a strong emotional impact on the visitors of the exhibition. Artistic arousal of emotions has been a traditional

Matteo Baldoni · Cristina Baroglio · Viviana Patti · Claudio Schifanella
Dipartimento di Informatica, Università degli Studi di Torino
e-mail: {baldoni,baroglio,patti,schi}@di.unito.it

C. Lai et al. (Eds.): New Challenges in Distributed Inf. Filtering and Retrieval, SCI 439, pp. 131–149.
springerlink.com
© Springer-Verlag Berlin Heidelberg 2013

matter of speculation in aesthetics from Aristotle onwards. Even though the matter intrinsically stimulates challenging issues, the most recent studies in analytic philosophy seem to converge on considering emotions in the aesthetic experience as an essential dimension of cognition, in other words as a means of discerning what properties an artwork has and expresses [13]. For what concerns the Art World, recently, emotions and sentiment monitoring have been the core of some important projects, conceived from both *pioneering artists*, see, for instance, the ongoing series on the mechanics of emotions by the contemporary new-media artist Maurice Benayoun[1], and from *cultural institutions* and *museums*. Think, for example, to the Swiss national five-years long research project "e-motion: Mapping Museum Experience" [29], where the psycho-geographical effect of the museum on the visitor were investigated and visitors' art reception as well as their psychological reaction were monitored by means of techniques for visitor tracking and of biometric measurements.

Moreover, it is widely recognized that technology has the power to transform visitors' experiences at museums. In particular, the Social Web and Web 2.0 technologies can play an important role in promoting the visitors participation, with a focus on networking individual visitor experiences [23]. Many museums and artistic associations opened their collections for access on the web and have studied the potential of social tagging [28, 5, 3]. According to the Social Web principles, users can be involved in the production of contents or in their elaboration, e.g. by publishing and organizing own materials, by posting comments for discussing published contents, by participating into wikis, by rating resources. Social networks and platforms promote the participation of users in many ways, by stimulating the expression of opinions about the contents inserted by other users, by supplying simple "Like" or "Dislike" tools, star-rating systems, tag-based annotation and navigation, and so forth. This huge amount of data is a precious information source about *perceptions, trends, emotions and feelings*.

One of the emerging research fields, targeted at extracting information from the data that is supplied by the Social Web users, is *emotion-oriented computing* (a.k.a. *Affective Computing* [20]), whose focus is to automatically recognize the users' emotions by analyzing their tagging or writing behavior. In particular the rise of social media has fueled interest in *Sentiment Analysis* (and *Opinion Mining*) [17, 7]: since emotions are often related to appreciation, knowing the feelings of the users towards target issues is an important feedback that can support many decisional tasks and has found interesting applications in the business world.

What we propose in this paper is to study how to fruitfully apply Sentiment Analysis to the Planet Art, by exploiting, as an information source, the tags that the visitors leave for commenting artworks. Tags can be left both in the case the collection is accessible only through the web and in the case an exhibition supplies to its visitors appropriate tools for playing with the (physical, in this case) artworks.

---

[1] http://www.benayoun.com/

In order to elicit emotional meanings we developed *ArsEmotica*, a prototype application, that extracts a shared emotional semantic from the floating individual perceptions and reactions of the visitors of a collection. ArsEmotica creates a semantic social space where artworks can be dynamically organized according to an ontology of emotions based on the one in [12], which includes the following base emotions: Anger, Fear, Happiness, Sadness and Surprise. It can be interfaced with a resource sharing and tagging system, which provides the data that is to be processed by the so called *emotional engine*. The emotional engine task is to automatically identify the emotions which better capture the affective meaning, that visitors collectively (though indirectly, through the tags as we will see) gave to the artworks.

We validated ArsEmotica by means of a *user study* which involved the community of Italian users of the *ArsMeteo* art portal (http://www.arsmeteo.org [1]). ArsMeteo is a Web 2.0 platform (on-line since 2006), that enables the collection and the presentation of digital (or digitalized) artworks and performances (including poems, videos, pictures and musical compositions), together with their tags. The user study focussed on a corpus of about 40 tagged artworks and involved about one hundred users (about 35% of the whole community), the same community which tagged the artworks. Notice that the tagging activity, monitored in ArsMeteo since 2006, was not performed with the aim of later applying some kind of Sentiment Analysis, but in a pure Web 2.0 approach, as a form of spontaneous annotation produced by the members of the community on one another's works. Moreover, in ArsMeteo, artworks usually have many tags, expressing a variety of meanings, thus supporting the emergence of different emotional potentials. This is consistent with the idea that art can emotionally affect people in different ways. When this happens, the analysis performed by ArsEmotica is capable of providing multiple emotional classifications.

The ontology of emotions that we used is based on the one in [12], and was constructed by enriching, with the help of ArsEmotica, the latter by adding over four hundred Italian words bearing an affective value. The correlation between tags and emotional concepts was computed by following the approach in [4], exploiting and combining Semantic Web tools and lexical resources. In particular, we relied on multi-lingual and Italian computational lexicons [19] and affective lexicons (WordNet-Affect [25] and SentiWordNet [10]).

Our approach to Sentiment Analysis, applied to the art domain, presents some novelties w.r.t. the literature. First, by relying on an *ontology*, we can produce results that are much richer than the classical polarized appreciation; this is necessary when dealing with the Art World, where reducing emotional evaluations to positive (negative or neutral) classifications is an inadmissible simplification. Second, we focus on a very essential form of textual resource: *collections of tags*. This choice is motivated by the fact that tags are one of the most common ways for annotating resources, and that by concentrating on tags, it is possible to leave aside the complications due to text analysis (often aimed at identifying the keywords inside text) and focus on the emotional analysis.

The paper begins with a brief overview of the background with a special focus on the art application domain and on visitor's sentiment. Section 3 describes the ArsMeteo tagging platform. Section 4 briefly presents the current version of the ArsEmotica prototype application. Section 5 reports and discusses the results of the user study. Final remarks end the paper.

2 Art, Sentiment and Social Media

Over the last five years, social media, by promoting user participation and peer-to-peer interaction, have transformed the ways that web users interact with content and with each other on the web. This affected and influenced also the "Planet Art". Many curators, cultural organizations, institutions, and museums started to explore the ways social technologies and principles of Web 2.0 can be applied in museums or exhibitions, with the main aim of encouraging an active discourse among visitors, and of fostering social awareness and reflections about collections of artworks. Experiments, like the one described in [6], proved that people value connections with other persons, both for their own sake and as a way for supporting a meaningful experience.

Along this line, some important museums (among which the American museums: Guggenheim, Metropolitan Museum of Art, and San Francisco Museum of Modern Art) opened their collections for access through the web [28, 5], and studied the potential of social tagging in the development of better interfaces, by supporting the *steve.museum* research project (2006-2008) [28]. Tagging, due to its highly subjective nature, is perceived by museums as a valuable feedback that reveals the way in which their public perceives collections, exhibitions, artworks.

The curator Nina Simon in the recent book "The Partecipatory Museum" [23] goes one step beyond by investigating the important role that artworks can play as "social objects", i.e. as a basis for an object-centered sociality [21], and in the foundation of a new generation of social applications. The key aim is to stimulate participation and encourage visitors to share their experiences with one another. Artworks, indeed, are a good example of "social objects" because they are objects that connect people, for instance by fostering conversation. They do so because they raise personal feelings and personal degrees of understanding and of acceptance. As a consequence, people can attach to them stories that are taken from their own experience. Artworks occupy a space, either physical or a piece of display, and this allows people to create a contact with them. They can be provocative.

The kind of work we are doing with ArsEmotica, on the collection provided by the ArsMeteo portal, finds its setting this picture but the emotional dimension moves to the foreground. In the aestetic experience, emotions are an important element of the cognition. The work of art is apprehended through the feelings as well through the senses [13]. ArsEmotica enhances the social value of the artworks by exploring the possibility of capturing their emotional value for a group of people, namely the persons who tagged the digitalized reproductions of the artworks in the collection. The tagging platform can be seen as a mediator artifact, allowing the ArsMeteo users to interact with the social objects. ArsEmotica, the tool that we developed for analysing the tags and for extracting the prevalent emotions associated to the artworks, can be seen as an interpretative tool which produces new information by combining the individual contributions of the users.

Having the emotional dimension in the foreground can be a key element for encouraging visitors of collections to share their experiences about the artworks as social objects. The outcome of the collective experience can be presented to the society of users into several forms. For instance, in the case of digitalized collections, by allowing the construction of personalized emotion-driven visits, while in the case of physical exhibitions by supplying an emotional map of the rooms. Whatever the tool used to display the outcomes of the emotional analysis, this will be the result of a co-creation process which involved all of the users who tagged the artworks.

This way of interacting with artworks is, to the best of our knowledge, original. In fact, if it is possible to find studies on mood-driven music selection [9] (see also applications like StereoMood and SourceTone), little has been done for visual artworks, like images, paintings, and photographs. Social platforms like Flickr consider artwokrs as social objects, placing them in the middle of the user's activity, but the way in which users interact with photographs has nothing to do with the inhancement of the emotional dimension: users can leave comments to a picture or annotations directly inside the picture, and they can use tags, but this huge amount of information is currently not used to convey an emotional meaning. In this respect, one partly exceptional application is the flash website Moodstream [15], where the user chooses a mood and the system provides appropriate images and a soundtrack from a corpus of pre-classified data.

A high interest in monitoring the sentiment of the visitors in environments like museums and exhibitions is witnessed by projects like *e-motion* [29][2], where the psycho-geographical effect of the museum on the museum visitors is under investigation by means of methods for visitor tracking and biometric measurements. Our current work with ArsEmotica attacks a sibling issue, i.e. to monitor the visitor sentiment by relying on semantic *social web technologies* (ontologies, folksonomies, tagging), and proposes an approach based on ontology-driven sentiment analysis. To the best of our knowledge, the proposal to apply Sentiment Analysis to the Planet Art is original.

Before describing our approach, let us briefly overview the background concerning the techniques and topics involved in the ArsEmotica proposal. Ontologies and folksonomies are two ways for indexing resources: the former are to be designed by knowledge engineers, while the latter are spontaneously produced by the *tagging* activity of the members of a community. Tagging is one of the ways offered to users by the so called social web to become actively involved into the web experience, and amounts to attaching freely chosen labels to resources. Often such labels are used to categorize the resources, but, especially in art portals like ArsMeteo, or in artistic domains, where resources represent artworks, movies, music, books, they are also

---

[2] Mapping Museum Experience:
http://www.mapping-museum-experience.com/en

used to express reception, opinions, feelings, and users seem to use tags to supply concise reviews.

Whatever they represent, tags are used as meta-data on top of which it is possible to devise many algorithms for navigation and retrieval. More precisely, these algorithms exploit the folksonomy, arising from tags, using it as an open, distributed and social classification system, though with a *flat* structure. This final remark suggests one of most challenging tasks that are currently being studied in the field, which is the identification of relationships that tie the folksonomy terms with one another, and possibly capture a machine-processable semantics. There are many attempts to reconcile folksonomies with ontologies, for instance by inducing ontologies from folksonomies or by matching terms in some way. It is out of the scope of this paper to get into the details. We would just like to mention some surveys by [8, 27], together with a previous work of the authors [3], where the association was done based on the outcomes returned by a search engine. The reason for these interests is that, indeed, the Social Web aims at developing applications that combine the ease of use, which is typical of its platforms, with the advantages deriving from a formal semantics, i.e. interoperability, data/service integration, personalization, better recommendation and retrieval performances [22].

In this context, the identification of the feelings of a community or of its single members is receiving an increasing attention, as an indicator of the appraisal of topics, people, situations, resources, trends. Hence the development of Opinion Mining and Sentiment Analysis [17, 7], of ontologies of emotions, like the one we started out work from [11], and of W3C markup language proposals [2]. However, there are still few applications that use the most advanced results in Semantic Web technology to deal with emotions [14] and most of the approaches use ontologies where emotions are individual, isolated units (e.g. WordAffect). Such considerations motivated our focus on the ontology of emotions in [11], an OWL ontology where emotions are structured and organized in levels, trying to integrate the results of the most recent psychological models. OntoEmotion provided a good starting point to explore an ontology-driven approach to Sentiment Analysis, where tags (and then tagged resources) are related to emotions. Such approach is new w.r.t. previous work on Sentiment Analysis and allows to extract from tags affective information which is richer than a polarized appreciation.

3 Arsmeteo

ArsMeteo (http://www.arsmeteo.org, see Figure 1 for a snapshot) is an art portal for sharing artworks and their emerging, connective meanings. It is inspired by an idea of the artist *Giorgio Vaccarino* and opens with a planetary vision of the Earth flown over by evolving clouds of words and images. The main aim of the project is to create a space for a community of artists and art lovers, where artists can find a dynamic, interactive and fertile background for *artistic experimentation* and *cooperative artistic creation*. Without any intermediation, artists can interact on the same artwork, simultaneously: thus, different ages, backgrounds, languages, ways of life can merge and generate a new level of composite culture. The web platform combines social tagging and tag-based browsing technology with functionalities for collecting, accessing and presenting works of art together with their meanings. It enables the collection of digital (or digitalized) artworks and performances, belonging to a variety of artistic forms including poems, videos, pictures and musical compositions. Meanings are given by the tagging activity of the community. All contents are accessible as "digital commons".

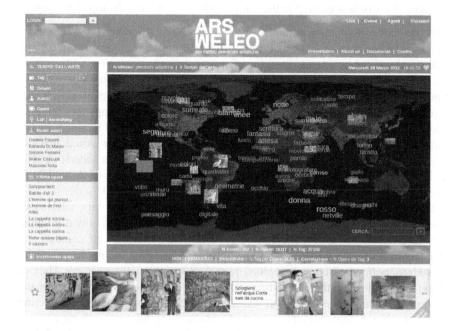

Fig. 1 The ArsMeteo portal entrance.

Artists and visitors may express their own reception of the artworks by tagging them or by clicking on the plus and minus symbols next to tags to change tag weights, which will affect the ranking of search results. Currently, the portal gathers a collection of about 9936 artifacts produced by 302 artists, it receives about 300 hits per day, and it has collected almost 37000 tags (an average of 13 tags per artwork). The tagging activity of the community provides a basis for browsing the artworks in traditional ways as well as according to different navigation methaphors, respectively called *Art Time Machine*, *Argonaut* and *Serendipity*:

- *The Art Time Machine*: the whole ArsMeteo archive, including forum posts and events, can be accessed by year;
- *Serendipity*: users can find resources *without* performing an explicit and systematic search. When a user chooses a graphical artifact, the starting point of the

search, the system puts a preview of such resource at the center of the search page. Then, the artwork is encircled by the previews of 24 graphical resources randomly chosen among all those which are tag-related with the initial choice. The user can then continue to browse by clicking on one of the surrounding resources: the serendipity game restarts by putting the new resource at the center of the page;

- *Argonaut* The idea is to visualize the evocative power of (poetic) sentences and text. The user starts by writing a sentence on the proper text area. Artworks tagged by words occurring in the sentence drop down from the top of the page.

This dynamic environment naturally engenders constructive contexts, where users can invent new forms of trans-individual artistic actions and reflections. In the last year we already observed the emerging of *artistic group actions*. New artifacts have been created by collecting, recycling and recontextualizing fragments of artworks from different users and from different media or instances. Different users synchronized their creative potential and generated new artworks and languages building upon (or reinventing) contents: new poems were created as rivers of words by the simultaneous and interactive tagging activity of many users; a new visual alphabet (*Alfameteo*) emerged by reinterpreting some artworks as alphabetic letters; in "The table of Niépce" series, authors started to play, again by using digital technologies, an artistic game of collective painting, which was invented in 1979.

For what concerns the *community aspects*, Arsmeteo started in a national context: most of the users are Italian, contemporary artists. Since the community is growing very fast, in order to give it the time to get acquainted with the new contents and give feedbacks by tagging (keep the descriptivity index high), we implemented the *SlowMeteo policy*: artists cannot upload more than 3 new resources per day. The risk of too many uploads is that interesting artworks can be unfairly neglected. For many of the ArsMeteo authors the portal was a first appealing opportunity for accessing and exploiting the new social potential of web-based technologies. Some of them entered in this new world thanks to the help of other more technologically skillful users, who play the role of *digital curators*.

4 ArsEmotica

This section describes the architecture of *ArsEmotica*, the application software that we developed. The analysis steps that we are about to describe rely on a pre-processing phase in which tags are filtered so as to eliminate flaws like spelling mistakes, badly accented characters, and so forth.

Figure 2 reports the three main steps that characterize the computation after the pre-processing:

- **Checking tags against the ontology of emotions.** This step checks whether a tag belongs to the ontology of emotions. Tags belonging to the ontology are immediately classified as "emotional".

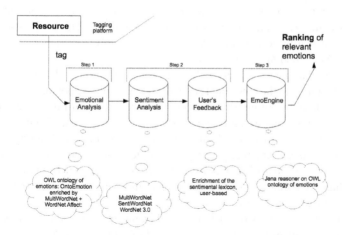

Fig. 2 ArsEmotica overall architecture.

- **Checking tags with SentiWordNet.** Tags that do not correspond to terms in the ontology are further analyzed by means of *SentiWordNet*, in order to distinguish *objective* tags, which do not bear an emotional meaning, from *subjective* and, therefore, affective tags. The latter will be the only ones presented to the user in order to get a feedback on which emotional concept they deliver.
- **Ranking of Emotions.** Based on data collected in the previous steps, the tool ranks the emotions associated by the users to the resource.

The following sections explain how the extraction of an emotional semantics is performed. We get into deeper details only for the first step of the algorithm, because the experimental results that we report and comment are aimed at validating the outcomes of this phase. Notice that for allowing the application even of the first step of the algorithm, we had to enrich the reference ontology of emotions with a suitable Italian vocabulary. The way we did this is also explained below.

4.1   The Ontology of Emotions and the Italian Emotional Words

The first step checks if the tags of a given resource are "emotion-denoting" words directly referring to some emotional categories of the ontology. Our starting point was *OntoEmotion*, an emotional ontology developed at Universidad Complutense de Madrid [11], that met our requirement to have a taxonomic structure, mirroring well-founded psychological models of emotions, and that was implemented by using semantic web technologies. The ontology is written in OWL and structures emotional categories in a taxonomy, which includes 87 emotional concepts.

The basic emotions are *Sadness, Happiness, Surprise, Fear* and *Anger* and the taxonomic structure basically refers to the psychological model by Parrot [18], adapted to these five basic emotions, and integrated with emotions which appear in other well-established models.

OntoEmotions has been conceived for categorizing emotion-denoting words. Classes corresponding to the emotional concepts were originally populated by about 250 instances, consisting of emotion-denoting words both from English and from Spanish. The ontology has two root concepts: *Emotion* and *Word. Emotion* is the root for all the emotional concepts. *Word* is the root for the emotion-denoting words, i.e. those words which each language provides for denoting emotions. Originally it had two subclasses: *EnglishWord* and *SpanishWord*. Each instance of these two latter concepts has two parents: one is a concept from the *Emotion* hierarchy (the class of emotion denoted by the word), while the other is a concept from the *Word* hierarchy (e.g. the language the word belongs to). For instance, the word *rage* is both an instance of the concept *Fury*, and an instance of the concept *EnglishWord*, which means that *rage* is an English word for denoting fury.

Since the tags used in our case study are mainly Italian words, we enhanced the ontology by adding the new subclass *ItalianWord* to the root concept *Word*, and semi-automatically populated the ontology. The approach we applied relies on the use of the multilingual lexical database MultiWordNet, in which the Italian WordNet is strictly aligned with Princeton WordNet 1.6., and its affective domain WordNet-Affect, a well-known lexical resource that contains information about the emotions that the words convey. A human expert checked the identified terms. WordNet is a lexical database, in which nouns, verbs, adjectives and adverbs (lemmas) are organized into sets of synonyms (synsets), representing lexical concepts. After choosing the representative Italian emotional words for each concept, such words were used as entry lemmas for querying the lexical database. The result for a word is a synset, representing the "senses" of that word, which are labeled by MultiWordNet unique synset identifiers. Each synset was then processed by using WordNet-Affect: when a synset is annotated as representing affective information, then, *all the synonyms belonging to that synset are imported in the ontology as relevant Italian emotion-denoting words*. This allowed us to automatically enrich the ontology with synonyms of the representative emotional words, but also to filter out synsets which do not convey affective information.

Let us see an example. When we query the MultiWordNet database with the italian word *panico* (noun, representative for the emotion *Panic*), only two of the three resulting synsets are affective (WordNet senses n#10337390 and n#05591377). In particular, the third not affective synset refers to the sense of the word "panico" described by the following gloss: *coarse drought-resistant annual grass grown for grain, hay and forage in Europe and Asia and chiefly for forage and hay in United States.* Thanks to our affective filter we can exclude words belonging to that synset (Setaria_italica, pabbio_coltivato) when populating the concept *Panic* of our ontology. The resulting ontology contains more than *450 Italian words* referring to the 87 emotional categories of OntoEmotion. In order to keep track in the ontology of the synonymy among words belonging to a same synset, we have defined the OWL object property *hasSynonym*.

ArsEmotica uses the enhanced ontology for checking if a tag describing a resource *directly* refers to some emotional category (Emotional Analysis). If yes, the tag is immediately classified as "emotional". The information collected during this phase is stored as a set of triples having the form: (t, e, s), meaning that tag t is related to emotion e with a strength value s. In general, as we will see in the remainder of the section, the range of the score s is $[0, 100]$ but when a tag is an instance of an emotional concept, its strength will be 100. So, for example, since the word "affanno" (breathlessness) is an instance of "anxiety", the corresponding triple will be: ("affanno", "anxiety", 100). Triples are store in a data base table.

4.2 Sentiment Analysis and User Feedback

The previous analysis identifies a set of tags as directly bearing an affective meaning. However, other tags can potentially convey affective meaning and *indirectly* refer to emotional categories of the ontology. We only sketch, here, the way in which we identify the tags which indirectly bear an affective meaning. The interested reader can find details in [4].

As observed in [26], some words can be emotional for someone due to her individual story. In other cases the affective power is part of the collective imagination (e.g. words like "war"). As a consequence, it seems appropriate and promising to involve the community in the definition of such indirect affective meanings. In order to minimize the effort requested to the users, before offering the tags to their judgment, we select the most promising ones by using SentiWordNet 3.0, a lexical resource for opinion mining where synsets of Princeton WordNet 3.0 are annotated according to their degree of neutrality, positiveness and negativity. Each synset s is associated the scores $Pos(s)$, $Neg(s)$ and $Obj(s)$ indicating how neutral (Obj) or affective (Pos and Neg) the terms contained in the synset are. Each score ranges in $[0.0, 1.0]$ and their sum is 1.0 for each synset. Since SentiWordNet was created for the English language, we needed to use MultiWordNet to align the Italian lemmas corresponding to the English ones. Moreover since SentiWordNet annotates a newer version of Princeton WordNet (3.0) with respect to the version MultiWordNet is based on (1.6), we have to query such newer lexical database.

The objectivity of a word in a given sense is simply measured as *1 - (Pos(s) + Neg(s))*. The value 1 indicates that the term is objective, while 0 means that the term conveys some strong sentimental (positive or negative) meaning. Different senses of the same term can have different opinion-related properties and different scores. When for no sense of a given term has a significant sentimental score, we conclude that it is mainly descriptive and usually does not evoke emotions. Therefore, we ask the evaluation of the community only for those terms having at least one meaning with a relevant sentimental score. This was done to have no false negative.

For all tags resulting potentially affective, users will be free to associate to the word one or more emotions from the emotional categories of the ontology with a strength value, which intuitively represents the user's measure of the semantic affinity of the term with the chosen emotional category. Again a set of triples (t,e,s) will be collected during this phase and stored in the data base.

4.3 Getting the Predominant Emotions

Once the analysis of the tags associated to a resource is finished, during the last step ArsEmotica ranks the emotions associated by the users to the resource and computes the prevalent emotion. Also this step is described in greater details in [4].

The ranking is performed with the help of the Jena Reasoner, applied to the triples collected during the previous analysis steps. The implemented algorithm relies on the taxonomic structure of the ontology and is inspired to the one in [12], where an analysis is performed to emotionally mark up a sentence by analyzing the words that compose it. Intuitively, the algorithm allows the selection of the *most specific emotion* which represents the affective information related to the artwork from the probability (score) that each of its tag has of indicating different emotions. The basic steps are: (1) processing the emotional concepts appearing in the triples (t,e,s), so as to identify also those emotions in the ontology that are related of the ones appearing in the triples. The identified emotional concepts can be organized into *layers* by following the parent-child relationship (in this phase a Jena Reasoner has been applied to the collected triples); (2) starting from the leaves and moving upward towards the root, compose and propagate the scores.

5 User Study and Evaluation

In order to test the effectiveness of the proposed system, we conducted a user study by involving the ArsMeteo community (285 users), i.e. the same community of artists who in part produced and in part tagged the artworks on which experiments were conducted. First of all, we selected a set $A = \{a_1,...,a_{38}\}$ of 38 artworks[3] from the ArsMeteo corpus, randomly selected among those which are the most voted by the community, explicitly including also artworks which provoke contradictory reception, e.g. artworks with multiple classifications, meaning that they are capable of dividing the community in its perceptions. Then, we used ArsEmotica to compute, for each artwork $a_i \in A$, the set $E_{a_i} = \{e_{a_i}^1,...,e_{a_i}^n\}$ of associated emotions, by analysing the tags attached to images by the ArsMeteo community. The number of emotions associated to each artwork in the corpus ranges from 1 to 9, with an average value of 2.9; the corpus generated 44 different emotions out of the set E of 87 emotional concepts, contained in the ontology (see Section 4).

Users were asked to answer a questionnaire composed of ten questions, belonging to two categories. In the first kind of question ($Q1$), starting from a randomly

[3] The whole set of artworks can be seen at http://di.unito.it/arsemocorpus

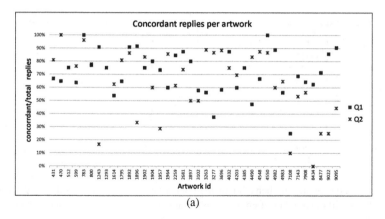

(a)

	Replies per artwork	
	Average concordant	Variance
Q1	71.43%	2.76%
Q2	64.04%	6.44%

(b)

Fig. 3 (a) The graph shows on the X-axis all artwork identifiers, on the Y-axis percentages of replies that are concordant to ArsEmotica outcomes; (b) The table summarizes the results achieved, by reporting the averages and variances of percentages for both kinds of questions.

Table 1 The table reports the total number of concordant replies, the total number of answers (both without dividing them per artwork) and the percentages of the former w.r.t. the latter.

	Overall Replies		
	#concordant	#total	%concordant
Q1	352	495	71.11%
Q2	347	495	70.10%

selected artwork $a_i \in A$, users were asked to choose the most representative emotion out of a set of two elements: one of the emotions that ArsEmotica identified as associated to a_i and one that resulted as not being associated to a_i. Formally, the user had to choose between $\{e_1, e_2\}$, where $e_1 \in E_{a_i}$ and $e_2 \in (E - E_{a_i})$. In the second kind of question ($Q2$), given a randomly selected emotion $e_j \in E$, users were asked to choose the most representative artwork between a set of two elements $\{a_1, a_2\}$, that were selected in such a way that $e_j \in E_{a_1}$ and $e_j \notin E_{a_2}$.

Among all, 99 ArsMeteo users replied to our call (about the 35%): each of them replied five randomly generated quiz both for question $Q1$ and for question $Q2$. For each type of question, we first analyzed the users' replies by measuring, for each artwork in A, the quantity of associations $(a_i, e_j), a_i \in A, e_j \in E$ between the artwork and emotions that are concordant to the output of ArsEmotica system w.r.t to whole set of replies. As can be seen in Figure 3(a), the majority of the artworks in the corpus show a correlation (corresponding to a high percentage of concordant replies)

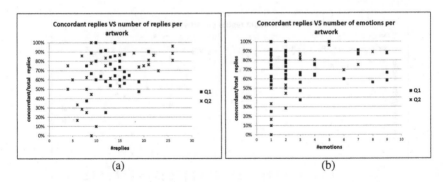

(a)   (b)

Fig. 4 The charts reports the behavior of the concordant replies w.r.t. (a) the total number of answers and (b) the number of emotions associated by ArsEmotica, for each artworks and kind of question.

between the emotions chosen by the users and those extracted by the ArsEmotica system. This behaviour can be observed both for questions $Q1$ and $Q2$, with values up to 100% for artworks 783, 4550, and 470. Figure 3(b) summarizes the results into a table, reporting the average value and the variance of all percentages of concordant replies obtained for each artwork. For what concerns question $Q1$, there is an average agreement with ArsEmotica up to about 71%, with a variance of 2.76%; the average for question $Q2$ decreases to 64% with a higher variance (over 6%), showing in any case a correlation between the user replies and the output of ArsEmotica. Table 1 reports the results without grouping the answers artwork per artwork. In this case, while for question $Q1$ the overall percentage of concordant replies is similar to what previously showed, this percentage increases to 70.11% for question $Q2$: this improvement is due to the fact that for this kind of question, poor results (for concordant replies) depend more on artworks, for which we gathered a low number of answers – indeed, due to random choice not all artworks were presented to users with the same frequency. This kind of artworks increase their weights when the overall percentage of concordant replies is computed by starting from artworks grouping. Figure 4(a) shows more details by positioning each artwork on a plane where the X-axis represents the number of replies it received and the Y-axis reports the percentage of concordant answers. A low number of answers necessarily turns out into a greater variance. Finally, in Figure 4(b) we analysed the relationship between the number of concordant replies and the number of emotions associated to an artwork by ArsEmotica: as one can see, in general, even though there is no sharp correlation, it is possible to notice that when the number of emotions increases, the number of artworks with low values of concordant replies decreases.

Some considerations it is interesting to draw, also for explaining variances, concern the sharpness of attribution of an artwork to one precise emotional concept, starting from tags. In general, one would think that when a tool like ArsEmotica identified one single emotion out of tags associated to an artwork, there will be not much surprise coming out from the users' evaluation. In some cases, however,

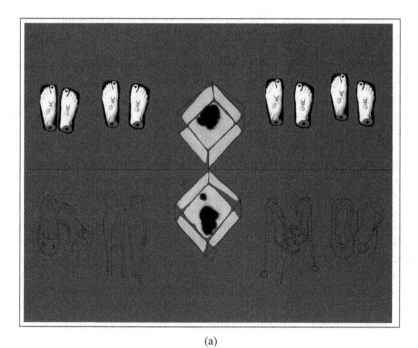

(a)

Tag			
	peli	rosso	desiderio
	feticismo	calzature	tacchi
	sangre	scatole	mozzato
	possesso	insiemi	smembrata
	segmenti	appiedata	sezionate

Per intervenire sui tag è necessario identificarsi.
Se desiderate un identificativo o lo avete dimenticato,
potete contattare mail@arsmeteo.org

Opera inserita il **8 Gennaio 2010** • vista **109** volte • con **6** interventi (l'ultimo il 3 Agosto 2011)

le opere di Elvira Biatta -

(b)

Fig. 5 (a) Artwork 8434: The Red Shoes, by Elvira Biatta; (b) The tags associated to artwork 8434.

it may happen (as for image 8434, see Figure 5) that an image which does not bear an emotional value is associated to some emotion because one of its tags has a meaning that can be related to an emotion. The fact that this correlation is almost accidental emerges from the users' evaluation: indeed, no user felt the emotion that ArsEmotica concluded image 8434 to convey, see Figure 3(a), even though in the repository there is a tag attached to it by a person, which is an emotion: desiderio (i.e. desire). This is an extreme case and, as one can see from the graph, there are not many other similar cases. However, they affect the results. Interestingly, by doing so, they give insight on the fact that in order to distinguish between accidental (or weak) emotional correlations and sharp but effective emotional correlations, it may be useful to introduce in the ArsEmotica algorithm a notion of strength of the emotion-bearing tags. We mean to explore this kind of enhancement in future work.

Another general observation is that it is quite unlikely for an artwork to be emotion-bearing for everybody and, when it is, to be related by everybody to a same single emotion. It is much more frequent the case in which there is a sharp classification towards a little set of emotions. This is coherent with the fact that tags are produced by a community, each of whose users relates to the artwork in a totally personal way, based on his/her personal background and story. However, as the results witness, it is possible to develop tools, like ArsEmotica, which are able to capture the main trends.

6   Conclusion and Future Work

With this work we have shown that it is possible to extract rich, emotional semantic information from the tags associated to a visual resource. The effort needed to obtain this outcome involves the combination of approaches, that were developed in different, though neighbouring, scientific disciplines, as well as of tools and technologies (e.g. lexicons and libraries) that are already available. The tool that we developed, ArsEmotica, succeeded in populating an ontology of emotions (based on [11]) with Italian emotion-bearing terms. Moreover, the first experiments proved that it is capable of extracting the prevalent emotions that are hidden inside the tags, associated to a digitalized visual artwork, a kind of response that is much more informative than the general polarized appreciation, returned by standard sentiment analysis tools. The analysis also gave interesting insights on images to which a single emotion seems to be associated, and images which seem to divide the visitors in groups with different perceptions.

We think that the proposed approach may be particularly suitable to those application domains where tags can be interpreted as *concise reviews* (e.g. artworks, books, movies). Notice that our emotional engine can in principle be interfaced, by developing a simple API, with any Web 2.0 platform (i.e. Flickr, Youtube or steve.museum) that shows the standard functionalities of a social resource sharing system, i.e. collecting, and presenting, browsing and accessing digital resources (artworks in our case) together with their tags. Given appropriate pre-processing tools capable to extract the relevant words from a text, its use could be extended also capturing the latent emotions behind textual comments.

The proposed solution can be refined in many ways. For what concerns the pre-processing of tags, we intend to improve the current prototype by applying stemming and word-similarity algorithms. For instance, in Italian, adjectives are declined in many ways, depending whether they refer to males or females, singular or plural. Stemming and lemmatization algorithms would help reduce the noise due to this variability. Word similarity could, instead, help to find relations among concepts that are not detected by the studied computational lexicons.

Moreover, the experiments that we carried on so far, only concern the first step of the ArsEmotica algorithm. Further extensive validation is to be performed also for the part described in Section 4.2. In this concern, the delicate point is the identification of a simple and effective way for motivating the users of ArsMeteo to annotate tags having an indirect affective meaning by means of emotional concepts, taken from the ontology. In order to face this issue, one promising direction could be to rely on the Game With A Purpose [30] paradigm and to develop a proper game in which users, as a side effect of playing, perform the task of associating emotional concepts to tag-words. This kind of solution would be along the line of recent approaches, which face the challenge of increasing the user involvement in building the Semantic Web [24]. An alternative could be to integrate in ArsEmotica the use of automatic techniques, e.g. the one proposed in [3], for identifying the association of terms having an emotional value (that is recognized by the sentiment analysis step) with the proper ontological concepts.

Moreover, it is known that the emotional semantics may vary depending on the context. Psychological theories concerning emotions, that tie perception to context, could be integrated in ArsEmotica to refine the outcome [16].

For what concerns the possible uses of ArsEmotica, we think that, by devising proper interpretative graphical representations for presenting the outcomes of the elaboration of the tags associated to artworks, it would be a precious co-creation instrument for museums and virtual galleries, along the direction traced in [23]. In this sense, it would help transforming classical art-fruition experiences into innovative, more immersive experiences, with a greater impact on visitors. Finally, the capability of extracting prevalent emotions foster the development of emotion-aware search engines and of emotional tag clouds. This would open the way to a plethora of applications, including smart-phone apps, not only with a cultural flavor (along the lines of the application in the previous section) but also more intrinsically related to leisure.

Acknowledgement. The authors thank all the persons who supported the work with comments and contributions. In particular, Paolo Rena, for the generous work on the ArsEmotica software development, Andrea Bolioli and *CELI s.r.l.* for the fruitful discussions on lexical and sentiment resources, Flavio Portis, Giorgio Vaccarino and the *Associazione Culturale ArsMeteo* for providing the data corpus, Fondazione Bruno Kessler and ISTI CNR for supplying the lexicons, the NIL group at the Complutense University of Madrid, which supplied *OntoEmotion*.

References

1. Acotto, E., Baldoni, M., Baroglio, C., Patti, V., Portis, F., Vaccarino, G.: Arsmeteo: artworks and tags floating over the planet art. In: Proc. of the 20th ACM Conference on Hypertext and Hypermedia, HT 2009, pp. 331–332. ACM (2009)
2. Baggia, P., Burkhardt, F., Oltramari, A., Pelachaud, C., Peter, C., Zovato, E.: Emotion markup language (emotionml) 1.0, http://www.w3.org/TR/emotionml/i
3. Baldoni, M., Baroglio, C., Horváth, A., Patti, V., Portis, F., Avilia, M., Grillo, P.: Folksonomies meet ontologies in arsmeteo: from social descriptions of artifacts to emotional concepts. In: Borgo, S., Lesmo, L. (eds.) Formal Ontologies Meet Industry, FOMI 2008, pp. 132–143. IOS Press (2008)
4. Baldoni, M., Baroglio, C., Patti, V., Rena, P.: From Tags to Emotions: Ontology-driven Sentiment Analysis in the Social Semantic Web. Intelligenza Artificiale: the International Journal of the AI*IA (to appear, 2012)
5. Chan, S.: Tagging and searching: serendipity and museum collection databases. In: Museums and the Web 2007, pp. 87–99 (2007)
6. Cosley, D., Lewenstein, J., Herman, A., Holloway, J., Baxter, J., Nomura, S., Boehner, K., Gay, G.: ArtLinks: fostering social awareness and reflection in museums. In: CHI 2008: Proc. of the 26th Annual SIGCHI Conference on Human Factors in Computing Systems, pp. 403–412. ACM, New York (2008)
7. Delmonte, R., Pallotta, V.: Opinion Mining and Sentiment Analysis Need Text Understanding. In: Pallotta, V., Soro, A., Vargiu, E. (eds.) DART 2011. SCI, vol. 361, pp. 81–95. Springer, Heidelberg (2011)
8. Dotsika, F.: Uniting formal and informal descriptive power: Reconciling ontologies with folksonomies. International Journal of Information Management, 407–415 (October 2009)
9. Dunker, P., Nowak, S., Begau, A., Lanz, C.: Content-based mood classification for photos and music: a generic multi-modal classification framework and evaluation approach. In: Proceedings of the 1st ACM International Conference on Multimedia Information Retrieval, MIR 2008, pp. 97–104. ACM, New York (2008)
10. Esuli, A., Baccianella, S., Sebastiani, F.: Sentiwordnet 3.0: An enhanced lexical resource for sentiment analysis and opinion mining. In: Proc. of the 7th Conference on International Language Resources and Evaluation (LREC 2010), ELRA (May 2010)
11. Francisco, V., Gervás, P., Peinado, F.: Ontological Reasoning to Configure Emotional Voice Synthesis. In: Marchiori, M., Pan, J.Z., de Sainte Marie, C. (eds.) RR 2007. LNCS, vol. 4524, pp. 88–102. Springer, Heidelberg (2007)
12. Francisco, V., Peinado, F., Hervás, R., Gervás, P.: Semantic Web Approaches to the Extraction and Representation of Emotions in Texts. NOVA Publishers (2010)
13. Goodman, N.: Languages of art: an approach to a theory of symbols. Hackett (1976)
14. Grassi, M.: Developing HEO Human Emotions Ontology. In: Fiérrez-Aguilar, J., Ortega-Garcia, J., Esposito, A., Drygajlo, A., Faúndez-Zanuy, M. (eds.) BioID MultiComm2009. LNCS, vol. 5707, pp. 244–251. Springer, Heidelberg (2009)
15. Images, G.: Moodstream (2010), http://moodstream.gettyimages.com/
16. Ogorek, J.R.: Normative picture categorization: Defining affective space in response to pictorial stimuli. In: Proc. of REU 2005 (2005)
17. Pang, B., Lee, L.: Opinion Mining and Sentiment Analysis (Foundations and Trends(R) in Information Retrieval). Now Publishers Inc. (2008)
18. Parrot, W.: Emotions in Social Psychology. Psychology Press, Philadelphia (2001)
19. Pianta, E., Bentivogli, L., Girardi, C.: Multiwordnet: developing an aligned multilingual database. In: Proc. of the First International Conference on Global WordNet (January 2002)

20. Picard, R.W.: Affective computing. Technical Report 321, MIT (1995)
21. Schatzki, T.R., Knorr-Cetina, K., Savigny, E.: The practice turn in contemporary theory, Routledge (2001)
22. Shadbolt, N., Berners-Lee, T., Hall, W.: The semantic web revisited. IEEE Intelligent Systems 21(3), 96–101 (2006)
23. Simon, N.: Participatory Museum. Museum 2.0 (2010)
24. Siorpaes, K., Hepp, M.: Games with a purpose for the semantic web. IEEE Intelligent Systems 23, 50–60 (2008)
25. Strapparava, C., Valitutti, A.: WordNet-Affect: an affective extension of WordNet. In: Proc. of LREC, vol. 4, pp. 1083–1086 (2004)
26. Strapparava, C., Valitutti, A., Stock, O.: The affective weight of lexicon. In: Proc. of LREC, pp. 1–83 (2006)
27. Trant, J.: Studying Social Tagging and Folksonomy: A Review and Framework. Journal of Digital Information 10(1) (2009)
28. Trant, J., Wyman, B.: Investigating social tagging and folksonomy in art museums with steve.museum. In: Proc. of the Collaborative Web Tagging Workshop, WWW 2006 (2006)
29. Tschacher, W., Greenwood, S., Kirchberg, V.: Physiological correlates of aesthetic perception in a museum. Psychology of Aesthetics, Creativity, and the Arts (2011)
30. von Ahn, L.: Games with a purpose. IEEE Computer 39(6), 92–94 (2006)

OntoTimeFL – A Formalism for Temporal Annotation and Reasoning for Natural Language Text

Francesco Mele and Antonio Sorgente

Abstract. This work has been developed with the motivation of defining a formalism (OntoTimeFL) for annotating complex events in natural language texts and applying, to the items annotated, several types of axioms and rules for temporal reasoning. In part, OntoTimeFL is a conceptualization of TimeML formalism, where the basic concepts of annotation by an ontological form have been represented. TimeML entities have been analyzed to extract the concepts hidden in the formalism and such concepts have been defined as classes of a formal ontology. In addition, OntoTimeFL introduces new constructs (concepts) for the annotation that mainly concern three complex events: narrative, intentional, and causal. The definition of OntoTimeFL through a formal ontology is a methodological choice made in order to facilitate the   automatic annotation processes, to reuse the existing axiomatics in the research of temporal reasoning, and to facilitate the creation of new ones (in particular an application of causal reasoning will be shown).

Keywords: Causal reasoning, Complex Event, Natural Language, Ontology, Frame Logic.

1 Introduction

This paper deals with problems inherent the representation, annotation, and temporal reasoning of events present in natural language texts. In particular, attention has been given to natural language texts that are present in a distributed context like Internet. These texts hide a multitude of events, that are connected by

Francesco Mele · Antonio Sorgente
Istituto di Cibernetica "E. Caianiello",
Consiglio Nazionale delle Ricerche
Via Campi Flegrei, 34 Pozzuoli (Napoli) Italia
e-mail: {f.mele,a.sorgente}@cib.na.cnr.it

C. Lai et al. (Eds.): New Challenges in Distributed Inf. Filtering and Retrieval, SCI 439, pp. 151–170.
springerlink.com   © Springer-Verlag Berlin Heidelberg 2013

extremely difficult to detect relationships, and where it is clear that certain events belong to an also hidden whole still to be discovered and explicitly represent. Attention has been focused on those aggregations of events that in a lot of research [19] have been labeled as complex events. Part of this work has been directed to the formalization of complex events present in the form of news, facts and stories on the Internet. Some concepts represented in OntoTimeFL emerge from TimeML, a standard mark-up language for the annotation of time, events, and relation between them in a natural language text.

Regarding news and stories, something extremely interesting and useful for potential applications is the study of the causality among events. Frequently, for a particular event in a story it is useful to know the events that have caused it. For the large number of events expressed in natural language on the Internet, one can provide advantageous solutions just by building adequate and efficient automatic tools. For this purpose, a logic program for causal reasoning has been defined and an example of application to a story annotated by OntoTimeFL has been shown.

The context of these distributed texts gives rise to another problem that stems from several sources that can exist for the same news, and which may contain contradictions. The latter are propagated in the annotations that can be performed on these texts. If multiple annotations are present, then an analysis of non-contradictory from a temporal point of view is necessary (the typical case being that for the same pair of events E1 and E2, E1 temporally follows E2 and it is also true that E2 temporally follows E1). Another problem that occurs in natural language texts in a distributed context is the connectivity. In fact, when one considers news or stories written by different authors, even before starting any analysis (including that of inconsistency) it is necessary to detect if there are events in a narration not connected to the rest of events.

Problems of the connectivity and consistency of events in natural language, and the study related to their causation, are important topics of theoretical and fundamental research. For temporal and causal reasoning, the intention in this work is to provide solutions (even if like simplifications of one complete theory) that are immediately applicable (without any additional algorithms) to the problems existing in the distributed domains.

Some objections could be raised from the point of view of the cost of the method presented, evidencing that it requires manual annotations of the text. There is certainty in this methodology because some of the algorithms for the automated annotation of events have already been produced in other research [15].

1.1   TimeML

The TimeML is a standard mark-up language for annotating events and temporal expressions in a document. TimeML resolves some problems regarding the annotation of events, including: how to represent the anchorage of events on the temporal axis, the relations in which events have with respect to one another, and the attempted detection of temporal references when the context of the temporal expressions is unspecified. Some research projects about several natural languages have been developed in which some wide data banks have been produced. There are four main data structures that are specified in TimeML: TIMEX3, EVENT, SIGNAL, and LINK.

The TIMEX3 is used to annotate temporal expressions such as times, dates, durations, etc. TIMEX3 tag includes a type and a value along with some other possible attributes and allows specification of a temporal anchor.

TimeML considers the events as situations that happen or occur. The EVENT tag is used to annotate events in a text. EVENT includes a class attribute with values such as occurrence, state, or reporting. In addition the EVENT tag includes information about part of speech, tense, aspect, modality, and polarity.

The SIGNAL tag is used to annotate sections of text, typically function words, that indicate how temporal objects are to be related to each other. The terms marked by SIGNAL constitute several types of linguistic elements: indicators of temporal relations such as temporal prepositions (e.g on, during) and other temporal connectives (e.g. when) and subordinators (e.g. if).

The LINK tags encode the various relations that exist between the elements of a document (events and temporal expressions). There are three types of LINK tags: TLINK (Temporal Link) annotates the temporal relationship between the events or between an event and a time; SLINK (Subordination Link) annotates subordination relationships between the events; ALINK (Aspectual Link) annotates the relationship between an aspectual event and its argument event.

1.2 The Approach for the Construction of the Representation of Events

For the objectives of this work, the formalism TimeML [4, 13, 14] has many points of incompleteness and inadequacy. Nevertheless, TimeML is used as a starting point for the construction of complex events in the ontology for natural language defined in this research. In this regard, the particular approach adopted has been highlighted. There is some research [12] that proposes a number of external reference ontologies in order to enrich the syntactic segmentation of texts in natural language. Instead, the proposed approach is based on the analysis of already existing TimeML text annotations, in order to have emerge event classes and relevant temporal relationships between such classes.

It is natural to compare this approach to the work of Hobbes and Pusterjovsky [9] (see Fig. 1) in which some TimeML structures with the classes of the Daml-Time ontology [9] were put into correspondence. However, in the Pustejovsky & Hobbes work only the temporal entities (instant and intervals) were considered, neglecting the important tag EVENT of TimeML.

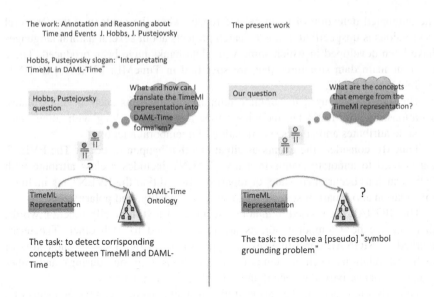

Fig. 1 Comparison between our approach and the approach defined in [9].

Like all formalisms expressed using XML [18], TimeML [4, 13, 14] has been affirmed both for the definition of its basic grammar, expressed by Document Type Definition (DTD), and for its easier to annotate main entities of natural language relative to time.

Unfortunately, the benefits that TimeML language offers in terms of ease of use, are matched by many disadvantages in terms of the ambiguity of the annotation operations and, especially, the extensibility of the language. In the annotation process with TimeML exist many entities of the natural language that do not have appropriate annotation tags. Also, among those same tags, are hidden concepts that need to be explicit (eg. the causal relationships between events), without being able to transfer results and axiomatics to annotated elements. An important example, in TimeML's tags there is no explicit notion of action.

The method of analysis adopted uses Formal Concept Analysis (FCA) techniques [8], in order to identify some implicit classes present in TimeML annotations.

OntoTimeFL can be considered as a conceptualization of the existing TimeML concepts, and also an extension of such formalism, in fact new concepts relating to the complex events and the causal relationships have been defined.

1.3 OntoTimeML Formalism for Event Reasoning Tasks for Natural Language Texts

OntoTimeFL permits for the annotation of all types of events derived from TimeML as well as those of complex events (narrative, causal, and mental events).

For this reason, a software module has been developed, which takes existing annotations in TimeML and translates them into OntoTimeFL (module (3) of Fig. 2). This program has been inserted in order to render compliant OntoTimeFL with TimeML and recover the wide set of annotations in TimeML.

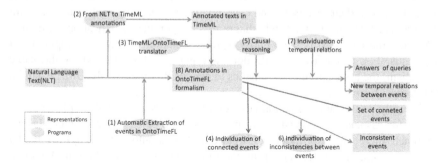

Fig. 2 The role of OntoTimeFL in event reasoning tasks for natural language texts.

OntoTimeFL allows to use several temporal reasoning axiomatics to the main entity of the representation (the events), regarding:

1. the correctness of the checking of an annotation;
2. the consistency of the checking of temporal relations;
3. the construction of rules for learning temporal relations between events;
4. the connectivity checking of the events of a narrative; and,
5. the use of causal relationships and their underlying reasons for the enrichment of temporal relations.

For temporal reasoning, a representation has been adopted that can be used by an axiomatic built for reasoning on temporal intervals and temporal instants.

The diagram (see Fig. 2) reports (with the exception of task 1 and 2) the modules implemented in our research. In this paper, the OntoTimeFL formalism (8) and the Casual Reasoning module (5) have been presented. Starting from the OntoTimeFL formalism it is possible to associate several axiomatics regarding: the detection of inconsistencies in events (6), the discovery of new temporal relations (7), and the analysis of their connectivity (4) (for more details about module (4), (6), and (7) see reference [11]).

This research has as its goal the creation of an architecture that allows, in a completely automatic way, the processing of text in natural language (eventually) through TimeML annotations, and the application of complex functionalities for temporal reasoning. For the module (2), an automated annotation that performs partial annotation (annotated events only) in TimeML of natural language texts [15] was produced.

1.4 Related Works

An important research project relating to the annotation of linguistic expressions is LMF (Lexical Markup Framework) [10], by which structures of the lexicon can be defined, or more precisely: LMF represents linguistic entities and their usage, through classes, their instances, and class relations associating such structures with the English text and using UML diagrams [17] for the visualisation. Although UML has the *Class* as its main entity of representation, it is not a computational apparatus. For this reason, UML does not permit (directly, and in the same formalism) the association of axiomatics to classes. In other words, UML is not an ontological formalism.

Recently, in the latter direction new research has been initiated. In [3] a unified model for associating linguistic information to ontologies was presented. In this approach there are two models LingInfo and LexOnto. The LingInfo model provides a mechanism for modelling the linguistic structures, while the LexOnto model enables the representation of external linguistic structures (predicate-argument structures in the form of ontological terms) and their association with corresponding ontological elements. This work also provides some requirements for associating linguistic information with ontologies; one of these consists of requiring representations in which the Linguistic and Ontological Levels are separated. LexInfo is based on the LMF formalism [10].

In other recent research [19], there have been several proposals for representing events. The basic motivation of this research consists in the belief that the events can constitute an excellent *structure* for aggregating knowledge. The large quantity of data and knowledge (fragmented and unstructured) on the Internet makes this research very interesting. An emerging methodology for the representation of events knowledge distributed on the Internet has been named Event-Centric [19]. To represent events, some formalisms have been inspired from a model that has its roots in journalism. This model called "W's and one H" adopts six attributes for the representation of events: Who, When, Where, What, Why, and How. This work is based on the model "W's and one H".

2 Explicit Representation and Compositional Nature of Complex Events

One of the main goals of this research is to build a model and a formalism for the representation and reasoning of complex events in natural language. There are several types of complex events, and TimeML formalism captures implicitly some of them through relationships expressed by the tags SLINK and TLINK from which one can define intentional and conditional events. In this context, attention to the intentional, causal, and narrative entities present in a natural language text have been focused on. In Natural Language research, many theories concerning the events have been produced, in particular for the formalization of a narrative that has been considered as a set of entities (events) and a set of temporal order relations between these events. In Artificial Intelligence (AI) much research on Mental Attitudes (Desire, Belief, Intention) [6] has been developed. Furthermore, in AI, Logic, and Philosophy [16], causality has been studied as a relationship between events and a set of rules (axioms) that justify and/or predict the causal order between events. In OntoTimeFL intentions, causations, and narratives, have been represented as complex events. Those entities can be themselves components of other complex events. For this purpose, a representation of complex events has been defined, where each complex event has:

- an interval and a temporal modality (before, during, etc. with respect to another interval, typical of simple events);
- an explicit representation by an ontological formalization, and by specific rules of composition for each type of complex event (causal, mental, etc.); and,
- a specific axiomatic associated to each class.

The explicit representation of complex events is motivated by the fact that these events can be considered (computationally) in the same way as a simple event in whose calculation there are temporal order relations, causal relations, or axioms defined for existing high-level reasoning. In addition, the utility of having an explicit representation of complex events is in itself of a compositional nature. A complex event can be composed of multiple events. In the proposed formalization a complex narrative event (a story) can be constructed by a set of event components and a set of order relations between these events. In such a formulation, the complex event "story" happens at a time interval that is determined from the time intervals in which individual events occur. In OntoTimeFL, intentional events have also been defined. They have been represented by a mental attitude and the object of intention that can be a physical event or another intentional event, useful for the representation of beliefs about beliefs of other agents. Finally, the proposed formalism allows one to define causation as compositions of two events: an event "cause" and an event "effect". This composition is also an event. For this event (causal complex event) the happen interval emerges from the happen interval of the components, that is, from cause event and effect event.

The OntoTimeFL formalism permits the annotation of complex events such as:

(Io$_{a1}$ credevo$_{e1}$ che tu$_{a2}$ non avessi voglia$_{act2}$ di andare$_{e3}$ al cinema ecco perché ti ho invitato$_{e4}$ a cena)$_{e6}$

(I$_{a1}$ believed$_{e1}$ that you$_{a2}$ did not want$_{act2}$ to go$_{e3}$ to the cinema, that's why I invited$_{e4}$ you to dinner)e$_6$

Informally, OntoTimeFL defines a complex intentional event (Io$_{a1}$ credevo$_{e1}$ che tu$_{a2}$ non avessi voglia$_{act2}$ di andare$_{e3}$ al cinema)$_{e5}$ that is composed of a mental event e1, which has as its argument in another mental event act2 (volere), which itself has another argument event (e3). In addition, the event e5 is represented as an event that happens over time, and, therefore, is a component of the causal event e6 (characterized by the causal relation cause(e5,e4)).

Then, this representation, one can annotate causal relations and associate rules (axioms) for their causal reasoning. In OntoTimeFL, these axioms can be easily integrated with other axioms of temporal reasoning by a well-known rule: if cause(e1,e2) then precedes(e1,e2).

3 Determining an Ontology of Events from TimeML Annotations

In this section the basic concepts defined in OntoTimeFL and how some of these concepts emerge from TimeML have been presented. Analyses through FCA techniques on an extensive archive of TimeML annotations [5] have been performed. The Flora2[1][7] formalism has been used for representing the concepts of our methodology.

3.1 Instant and Interval Representations

A mixed representation of time (based on point and time intervals) has been adopted. In OntoTimeFL, all temporal entities are represented as classes (Fig. 3). TemporalTerms is the main class and has several specializations: date or partial date (DateValue), time instants or combinations of them with date (TimeValue), symbolic times (Symbolic), and time intervals (IntervalValue). In addition, there are classes for annotating temporal entities in the text (Fig. 3). The definition of these classes has been inspired by the formalism TimeML in particular by TIMEX3 tag. The Backus Normal Form (BNF) tag of TIMEX3 is the following:

```
attributes  ::= id anchor type [functionInDocument]
              [beginPoint] [endPoint] [quant] [freq]
              [temporalFunction] [mod] [anchorTimeID]
              (value|valueFromFunction)
```

As shown in the BNF, there are some optional attributes (attributes between brackets) such as beginPoint, endPoint, quant, and freq. These attributes, however, hide the problem of not applicable (N/A); in fact, the annotation of these attributes is not always allowed, because the applicability depends on the value assumed by the attribute type. In order to eliminate this problem, an analysis using FCA techniques has been performed with the aim of

---

[1] To make reading simpler, some key constructs of this language are informally reported here: X::Y (class X is a subclass of Y), X:Y (X is an instance of Y), X =>Y (X is an attribute of type Y), X->Y (Y is the value of the attribute X), X *=>Y (same as X =>Y, but it is also hereditable from its subclasses).

defining an adequate representation for TIMEX3. In the end, the formulation of a taxonomy for TIMEX3 through a set of annotations TIMEX3 (see Table 1) has been defined.

Table 1 Annotations in table form of the tag TIMEX3.

Id	type	beginPoint	endPoint	quant	freq	value
t1	Duration	t11	t12	n/a	n/a	P3d
t2	Time	n/a	n/a	n/a	n/a	T16:00
t3	Set	n/a	n/a	ogni	1W	P1W
t4	Date	n/a	n/a	n/a	n/a	02-12-08

Optional attributes beginPoint, endPoint, quant, and freq as shown in the above table, hide the problem of N/A. The instances of these attributes are restricted to the value of type. In the table, in correspondence with the value Duration of attribute type there is always assigned a value to the attributes endPoint and beginPoint, and there is never assigned a value to the quant or freq (they are not applicable). From this annotation, it emerges that beginPoint and endPoints are descriptors (attributes) of a specific concept (class) that can be called Duration. The same observations apply to the attributes quant and freq.

The TIMEX3 tag highlights the problem of N/A attributes and the existence of hidden classes. Thus, four subclasses of the TIMEX3 class have been identified. The root of the taxonomy is TimeX3, and its subclasses are: TIMEX3Data, TIMEX3Time, TIMEX3Set, and TIMEX3Duration. These classes are associated with the TemporalTerms class. In Fig. 3 the TIMEX3 taxonomy is shown.

Fig. 3 Taxonomy of temporal terms.

3.2 The Representation of Events

In this work, an ontology for complex events has been defined: OntoTimeFL. OntoTimeFL has an abstract superclass (AnythingInTime) common to all entities that happen over time. Two subclasses are specializations of AnythingInTime: Event, wich represents the class of simple events, and ComplexEvent, wich represents the class of complex events. A sketch representation is given in Fig. 4:

Fig. 4 Taxonomy of the classes of simple and complex events.

In Fig. 4, in brackets, the attributes that are inherited from their respective superclasses are reported. `AnyThingInTime` is an abstract class (without instances) which is the superclass of the concrete classes: `Event` and `ComplexEvent`. The latter classes are the key concepts of the formalism OntoTimeFL.

```
Event::AnyThingInTime.
ComplexEvent::AnyThingInTime.
AnyThingInTime[ hasWhen*=>When,
               hasParticipants*=>Participant].
Event[hasWhat*=> Action].
ComplexEvent[hasWhy(AnyThingInTime)*=>CausalRelation].
```

The `Event` class has the descriptor `hasWhat`, which is associated with the class `Action`. Generally, this class describes an action (which happens over time) that characterizes an event or describes a property that is true in a specific time interval. The `ComplexEvent` is an abstract class and represents complex events through a set of events (components) and of relations between events. Its subclasses (see par. 2.4), in accordance with the type of relationship that exists between components, have been classified. `ComplexEvent` also has the attribute `hasWhy` that describes the causal relations between events. Why is a relation between two events; for this reason, it can only be the attribute of a complex event.

3.3 The Descriptor When

In OntoTimeFL there is a particular structure: the class `When` (see Fig. 2), which describes when an event happens using the effective symbolic interval (ESI) in which the event happens, and a temporal modality of happening, described by one (or more) temporal order relations (before, after, during, etc.) between ESI and some temporal interval of reference (or also another event). These relations have the goal of anchoring an event to the chronological axis, or with another event through a temporal order relation. The approach requires, therefore, that when an event (simple or complex) is created, it automatically generates a type identifier ESI, represented by two attributes: `sti`, the (effective) symbolic time in which the event starts (or in which the property is true), and `stf`, the time in which the event ends. The choice of having an effective time when an event happens and a temporal modality of happening, is motivated by the fact that often the effective time in which an event happens is not known, but one can easily know one or more relations for it (modality of happening: after a time tx, dtx before a certain range, etc.). Thus, even if one does not exactly know the exact value of the start and/or end of an event, one can annotate (or, automatically discover) relationships with other time intervals, as soon as they become available. The class When has the following definition:

```
When[hasTemporalMode*=>TemporalRelation,
   hasSymbolicHappenInterval*=>SymbolicHappenInterval].
SymbolicHappenInterval[sti*=> SymbolicTime,
                       stf*=> SymbolicTime].
```

An example of the When's structure has been provided. How can one annotate the When of the event in the text "*The earthquake of 1980*"? One cannot know when the earthquake effectively started and finished, but only that it occurred in 1980. Omitting the annotation of earthquake event, one realizes the following instance of When:

```
w1:When[ hasSymbolicHappenInterval*->symt1
         hasTemporalMode*->trel1].
symt1:SymbolicHappenInterval[stf->t1,sti->t2].
time1:DateCalendar[year->1980].
trel1:DuringEI[hasEntity1->symt1, hasEntity2->time1].
```

4   Simple Events and Actions

In OntoTimeFL, the class Event represents simple events. This class inherits the attributes of the superclass AnyThingInTime and has the attribute hasWhat, which describes an action that happens or a property that is true in a temporal interval.

```
Event[hasWhat*=>Action].
```

The attribute hasWhat has values in the Action domain and it describes exactly what happens (action) or what is true (property) in a temporal interval. From an analysis of the EVENT tag, emerges the existence of subclasses of the Event class (Fig. 5): OccurrenceEvent, ReportingThingsEvent, StateEvent, and PerceptionThingsEvent. These subclasses are defined in correspondence with the values of the CLASS attribute of the EVENT tag. The other values of the CLASS attribute, such as I_ACTION, I_STATE, and ASPECTUAL, are not defined as simple events but as complex events, because they always have a relationship with other events through the relation ALINK or SLINK.

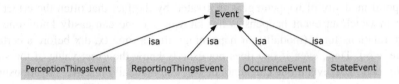

Fig. 5 Taxonomy of simple events.

The formalism has an explicit definition of the action concept (by the class `Action`). The subclasses of `Action` have been defined in accordance with the POS (part of speech): `AdjectiveAction`, `NounAction`, `PrepositionAction`, and `VerbAction`. Each subclass has subcategories inspired by TimeML formalism like `Reporting`, `Occurrence`, `Perception`, `State`, etc. The taxonomy of actions is emerged from FCA analysis performed on a set of annotated documents with TimeML (Table 2). Furthermore, Mental Acts (mental attitude as belief, desire, and intention) have been added compared to TimeML.

Table 2 Formal Concept Analyses (FCA) using TimeML annotations.

	POS (Part of Speech)					Class						
	Verb	Adjective	Noun	Preposition	Other	Reporting	Perception	Aspectual	I_Action	I_State	State	Occurance
1	X					X						
2			X			X						
3	X						X					
4	X							X				
5			X					X				
6	X								X			
7		X							X			
8			X						X			
9	X									X		
10			X							X		
11	X										X	
12		X									X	
13			X								X	
14				X							X	
15					X						X	
16	X											X
17		X										X
18			X									X
19					X							X

An example of OntoTimeFL annotation is the following:

Ho sonnecchiato$_{e1}$ un po'. (I dozed$_{e1}$ off a bit.)

```
act1:OccurenceVerb[annotatedWords->"Ho sonnecchiato",
        hasTense->PAST, has Aspect->PERFECTIVE,
        hasPolarity->POS].
e1:OccurenceEvent[hasWhat->act1, hasWhen->we1].
we1:When[ hasSymbolicHappenInterval->sIe1,
        hasTemporalMode->UNKNOWN].
sIe1:SymbolicHappenInterval[sti->st1,stf->st2].
st1:SymbolicTime.   st1:SymbolicTime.
```

5 Complex Events – Composition and Taxonomy

The ComplexEvent class is described by the method hasWhy(AnyThingInThing), a function, that given an input event belonging to a complex event, returns a set of causal relations that are the justification of why the event occurred.

ComplexEvent[hasWhy(AnyThingInThing)*=>CausalRelation].

Complex events are defined by a temporal mode, described by the descriptor hasWhen, the same attribute used for the description of simple events. For the simple event, the hasSymbolicHappenInterval term defines the temporal interval in which the action happens; while the time interval of occurrence of the complex event defines the minimum interval of time in which all events belong to the complex event that occurs. Therefore, the occurrence interval of complex events is not continuous, or likewise, not in all temporal subintervals is there an event that happens. The descriptor When can be calculated according to the descriptors of the events' components, or it is instantiated interactively. In the latter case, compatibility checks (defined by constraints), with respect to the attributes When of the events' components, must be run.

The TimeML allows for the annotation of one or more events connected through temporal relations, intentional relations, causal relations, etc. Although in TimeML there is no tag that allows the annotation of complex events, in this formalism exist relationships that implicitly *capture* a notion of a complex event. TLINK is an example of this relationship. In fact, it is a relation with two arguments E1 and E2 which presupposes there is another event E12 (complex event), having as its components (or parts) events E1 and E2. These events are related to each other by temporal order relations such as After, Before, etc.

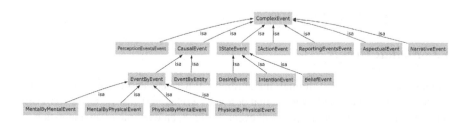

Fig. 6 The taxonomy (subclasses) of complex events.

Fig. 6 shows the taxonomy of complex events of the OntoTimeFL ontology, where the narrative events, the causal events, the intentional events (subclasses of IStateEvent), and the perceptual events have been labeled and represented as complex events.

5.1 Narrative Events

Complex narrative events (NarrativeEvent) define an aggregation of events connected by temporal relations. The components of NarrativeEvent can be simple events that describe actions that occur over time, or properties that are true in a temporal interval, or other complex events such as causal events (CausalEvent) or intentional events (IStateEvent).

NarrativeEvent, like all the subclasses of AnyThingInTime, inherits the attributes When and Participants, and like all the subclasses of ComplexEvent, inherits the method hasWhy(AnyThingInThing). The characterization of NarrativeEvent is given by the attribute hasTemporalRule, which defines temporal relations between components of the events. In fig. 7, an example of the annotation of the narrative event is given.

Fig. 7 An example of a narrative event.

5.2 Mental Events

Mental events are defined through the attributes When and Participants (inherited by AnyThingInTime) and by a slot that describes a relation (hasISateRelation) between a mental event and a physical event. In fig. 8, an example of an annotation of mental event is given.

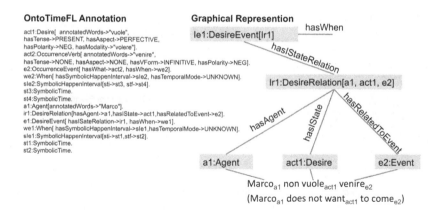

Fig. 8 An example of a mental event.

5.3 Causal Events

The complex event `CausalEvent` describes an event that relates to a cause-effect relation: the occurrence of an event (event cause) caused the occurrence of another event (event effect). Causal events are defined by a causal relationship between events, and like all event subclasses of `AnyThingInTime`, inherit the attributes When and Participant. In OntoTimeFL, a classification of causal events in accordance with the nature of the events involved has been defined, or, that is, if the cause-effect relationship is defined by physical and/or mental events.

The causal events `MentalByPhysicalEvent` are events formed by mental events (effects) and physical events (cause). The `PhysicalEvent` in these are perception action (see, look, smell, feel, etc.), while mental events are beliefs, desires and intentions. For example: *"He laughed and I thought he was joking"*.

The causal events `PhysicalByMentalEvent` are characterized by a Physical event that one thinks can be caused by a mental state of an agent. For example: *"I think it's a good book, I'll buy it"* and *"I would like something hot, I'll take a cup of tea"*.

The causal events `PhysicalByPhysicalEvent` are events composed by two events: the physical event cause and the physical event effect. For example: *"He bumped the glass with his elbow and broke it"* and *"It's raining and the road is wet"*.

The `MentalByMentalEvent` are events composed of two mental events: the mental event cause and the mental event effect. For example: *"I think it's the best team and I think it will win the championship"*.

In Fig. 9, an example of an annotation of `MentalByPhysicalEvent` is given.

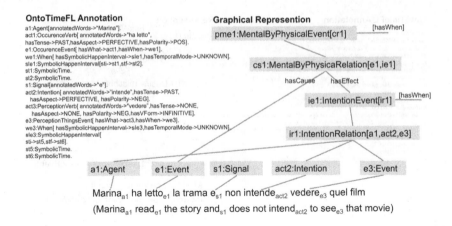

Fig. 9 Example of a `MentalByPhysicalEvent`.

Fig. 10 shows the graphical annotation of a `PhysicalByMentalEvent` event composed by mental events.

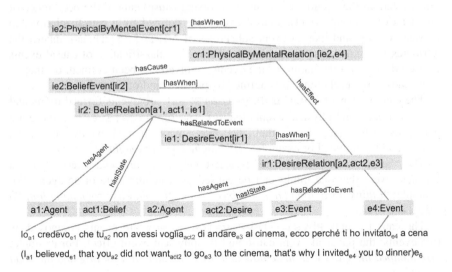

Fig. 10 Example of a `PhysicalByMentalEvent`.

For causal events, a widely shared relation (axiom), that brings together the causal relations with temporal relations, has been defined:

$$\text{BeforeEE[Ex, Ey]:-CausalRelation[Ex, Ey].} \tag{1}$$

If `Ex` is the cause of `Ey`, then the event `Ex` temporally precedes the event `Ey`.

6 An Axiomatic for Causal Reasoning

For causal reasoning, an axiomatic (a variant of the axiomatic defined in [2]) has been defined. The axioms that have been expressed (in a simplification of Flora2 [7], see note 2) are reported as follows:

```
Id1:CausalRelation[A, B]:-                    Strengthening
    Id2:BeforeEE[A, B], demo(A,B),
    Id3:CausalRelation[B,C], newId(Id1, Id2, Id3).
Id1:CausalRelation[A, C]:-                       Weakening
    Id2:CausalRelation[A,B], Id3:BeforeEE[B, C],
    demo(B,C), newId(Id1,Id2,Id3).
Id1:CausalRelation[A,B∧C]:-                          And
    Id2:CausalRelation[A,B], Id3:CausalRelation[A,C],
    newId(Id1,Id2,Id3).
Id1:CausalRelation[A∨B,C]:-                           Or
    Id2:CausalRelation[A,C], Id3:CausalRelation[B,C],
    newId(Id1,Id2,Id3).
Id1:CausalRelation[A, C]:-                           Cut
    Id2:CausalRelation[A,B], Id3:CausalRelation[A∧B,C],
    newId(Id1,Id2,Id3).
Id1:CausalRelation[A∧C,B]:-               Left Monotonicity
    Id2:CausalRelation[A, B], C:Event,
    Id3:BeforeEE[C,B], newId(Id1, Id2, Id3).
Id1:CausalRelation[A,B∨C]:-              Right Monotonicity
    Id2:CausalRelation[A,B], C:Event,
    Id3:BeforeEE[A,C], newId(Id1, Id2, Id3).
```

The predicate newId(Id1, Id2, Id3) generates a new id Id1 depending on Id2 and Id32.

In the axiom *Left Monotonicity*, the condition BeforeEE[C,B] has been included, because the event C is an event cause, and therefore to be added, it must precede B, otherwise, it generates a contradiction. For the axioms of *Weakening* and *Strengthening*, the meta-predicate demo(A,B) (implements the relation "B is deducible from A " [2]) has been defined, which was implemented as a variant of the meta-interpreter [1].

The axioms for causal relationships shown above have been defined for the class of causal events and can be applied to all subclasses of that class. In addition, the following corollaries, demonstrable from the axioms previously provided, have been defined:

```
Id1:CausalRelation[A,C]:-                   Transitivity
    Id2:CausalRelation[A,B],Id3:CausalRelation[B,C],
    newId(Id1, Id2, Id3).
Id1:CausalRelation[A ∧ B,C]:-               Substitution
    Id2:CausalRelation[A∧D,C],Id3:CausalRelation[B,D],
    newId(Id1, Id2, Id3).
```

The axioms, concerning the causal relationships described above, have been defined to be applied to the class of causal events and all subclasses of that class.

Using the formalism OntoTimeFL, an example of a text annotation, (from Wikipedia http://it.wikipedia.org/wiki/Storia_di_Napoli) on the history of Gioacchino Murat is shown. Causal reasoner, implemented in Flora2, starting from the axioms presented above, has been used to generate the simulation of the reasoning showed in Fig. 11. In Fig. 11, entities (events and relations) annotated by the user and those inferred by the reasoner are shown. Among the relations inferred there is `ce9:CausalRelation[e1∧a1,e5]` that claims the shooting of Murat (e5) was caused because of his (a1) having emitted a law (e3) and was ousted (e1).

Through reasoning then, one can build a chain of causes that constitutes the "Why" of events of a narrative.

One might consider *approximate* the causal implication "*tentò la,.., riconquista armata* "(*he attempted,...., reconquest of*) leading to " *fucilato* " (*shot*) (e5) of Murat.

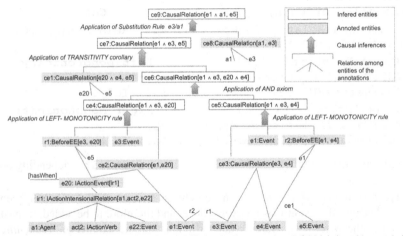

Fig. 11 A simulated example of causal reasoning after an OntoTimeFL annotation.

In order to refine the deductions, one could insert the most appropriate condition (implicit so not annotated) e30 *"Murat non rispettò la legge"* (*Murat did not respect the law*). Then, the annotation `ce1:CausalRelation[e30∧e4,e5]` is inserted.

That is, Murat was shot (e5) for the existence of the law (e4) and for not keeping to that law (e30). Then, the set of causal relations have to be enriched with the causal relation: `ce10:CausalRelation[e20,e30]` whose meaning is:

```
e20 ("tentò la,.., riconquista armata") cause
e30 ("Murat non rispettò la legge").

e20 ("he attempted,…, reconquest of") cause
e30 ("Murat did not respect the law").
```

Inserting the causal relationship `ce10` and applying the cut rule to the `ce10` annotation, one obtains, however, the relationship `ce1`. The inferential process then follows in the same way as the previous example (see Fig. 12).

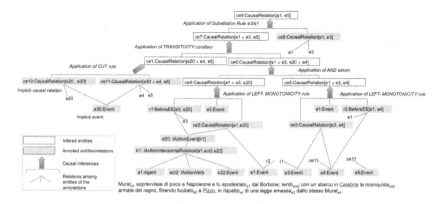

Fig. 12 The extension of the example presented in Fig. 11.

Conclusion

In this work, a formalism for the reasoning of complex events has been presented. Attention has been focused on three types of events: narrative, intentional, and causal. Starting from these three events, the formalism proposed (OntoTimeFL) permits for various compositions of a complex event which have a similar structure to a simple event (the descriptors When, What, etc). The only difference between simple events and complex events consists in the fact that the latter have attributes that emerge from the instances of the components. This paper has shown the possibility of associating some axiomatics to classes of complex events. In order to demonstrate this possibility, one detailed example of the causal reasoning of complex events about a cultural history has been provided.

Although the starting point of the analysis is the annotation of events in natural language texts, the intention of the authors is to provide results that can be used in other contexts where there are complex events.

References

1. Barták, R., Stepánek, P.: Extendible Meta-Interpreters. Journal Kybernetika 33(3), 291–310 (1977)
2. Bochman, A.: On disjunctive causal inference and indeterminism. In: Brewka, G., Peppas, P. (eds.) Proc. of the Workshop on Nonmonotonic Reasoning, Action and Change (NRAC), pp. 45–50 (2003)
3. Buitelaar, P., Cimiano, P., Haase, P., Sintek, M.: Towards Linguistically Grounded Ontologies. In: Aroyo, L., Traverso, P., Ciravegna, F., Cimiano, P., Heath, T., Hyvönen, E., Mizoguchi, R., Oren, E., Sabou, M., Simperl, E. (eds.) ESWC 2009. LNCS, vol. 5554, pp. 111–125. Springer, Heidelberg (2009)
4. Caselli, T.: It-TimeML: TimeML Annotation Scheme for Italian Version 1.3, Technical Report (2010)
5. TimeML Corpora, http://www.timeml.org/site/timebank/timebank.html
6. Cousins, S.B., Shoham, Y.: Logics of Mental Attitudes in AI. A Very Preliminary Survey. In: Lakemeyer, G., Nebel, B. (eds.) ECAI-WS 1992. LNCS, vol. 810, pp. 296–309. Springer, Heidelberg (1994)
7. Yang, G., Kifer, M., Wan, H., Zhao, C.: FLORA-2: An Object-Oriented Knowledge Base Language (2008), http://flora.sourceforge.net/
8. Ganter, B., Wille, R.: Formal Concept Analysis - Mathematical Foundation. Springer (1999)
9. Hobbs, J., Pustejovsky, J.: Annotating and Reasoning about Time and Events. In: Proceedings of AAAI Spring Symposium on Logical Formalizations of Commonsense Reasoning, Stanford, California (2003)
10. Lexical Markup Framework (LMF). ISO code number for LMF is ISO-24613:2008, http://www.lexicalmarkupframework.org/
11. Mele, F., Sorgente, A., Vettigli, G.: Designing and Building Multimedia Cultural Stories Using Concepts of Film Theories and Logic Programming. In: AAAI Fall Symposium Series 2010, Cognitive and Metacognitive Educational Systems (MCES) Symposium, pp. 57–63. AAAI Publications (2010)
12. Pazienza, M.T., Stellato, A.: The Protégé OntoLing Plugin: Linguistic Enrichment of Ontologies in the Semantic Web. In: Poster Proceedings of the 4th International Semantic Web Conference (ISWC 2005), Galway, Ireland (2005)
13. Pustejovsky, J., Castaño, J., Ingria, R., Saurí, R., Gaizauskas, R., Setzer, A., Katz, G.: TimeML: A Specification Language for Temporal and Event Expressions. Kluwer Academic Publishers, Netherlands (2003)
14. Pustejovsky, J., Lee, K., Harry, H.B., Boguraev, B., Ide, N.: Language Resource Management—Semantic Annotation Framework (SemAF)—Part 1: Time and events. International Organization (2008)
15. Robaldo, L., Caselli, T., Russo, I., Grella, M.: From Italian Text to TimeML Document via Dependency Parsing. In: Gelbukh, A. (ed.) CICLing 2011, Part II. LNCS, vol. 6609, pp. 177–187. Springer, Heidelberg (2011)
16. Tooley, M.: Time, Tense & Causation. Clarendon Press, Oxford (1997)
17. Unified Modeling Language (UML), http://www.uml.org/
18. Extensible Markup Language (XML), http://www.w3.org/XML/
19. Winkler, T., Artikis, A., Kompatsiaris, Y., Milonas, P.: Workshop Recognising and Tracking Events on the Web and in Real Life. CEUR Workshop Proceedings, vol. 624 (2010)

Representing Non Classical Concepts in Formal Ontologies: Prototypes and Exemplars

Marcello Frixione and Antonio Lieto

Abstract. Concept representation is still an open problem in the field of ontology engineering and, more in general, of knowledge representation. In particular, it still remains unsolved the problem of representing "non classical" concepts, i.e. concepts that cannot be defined in terms of necessary and sufficient conditions. In this chapter we review empirical evidence from cognitive psychology, which suggests that concept representation is not an unitary phenomenon. In particular, it seems that human beings employ both prototype and exemplar based representations in order to represent non classical concepts. We suggest that a similar, hybrid prototype-exemplar based approach could be useful also in the field of formal ontology technology.

Keywords: Concept Representation, Semantic Web, Semantics and Ontology Engineering.

1 Representing Non Classical Concepts

The representation of common sense concepts is still an open problem in ontology engineering and, more in general, in Knowledge Representation (KR) (see e.g. Frixione and Lieto in press). Cognitive Science showed the empirical inadequacy of the so-called "classical" theory of concepts, according to which concepts should be defined in terms of sets of necessary and sufficient conditions. Rather, Eleanor Rosch's experiments (Rosch 1975) – historically preceded by the philosophical

Marcello Frixione
University of Genova
e-mail: mfrixione@unisa.it

Antonio Lieto
University of Salerno
e-mail: alieto@unisa.it

C. Lai et al. (Eds.): New Challenges in Distributed Inf. Filtering and Retrieval, SCI 439, pp. 171–182.
springerlink.com   © Springer-Verlag Berlin Heidelberg 2013

analyses by Ludwig Wittgenstein (Wittgestein 1953) – showed that ordinary concepts can be characterized in terms of prototypical information.

These results influenced the early researches in knowledge representation: the KR practitioners initially tried to keep into account the suggestions coming from cognitive psychology, and designed artificial systems – such as frames (Minsky 1975) and early semantic networks (see e.g. Quillian 1968) – able to represent concepts in "non classical" (prototypical) terms (for early KR developments, see also the papers collected in Brachman and Levesque 1985).

However, these early systems lacked clear formal semantics and a satisfactory metathoretic account, and were later sacrificed in favour of a class of formalisms stemmed from the so-called structured inheritance semantic networks and the KL-ONE system (Brachman and Schmolze 1985). These formalisms are known today as *description logics* (DLs) (Baader et al. 2010). DLs are logical formalisms, which can be studied by means of traditional, rigorous metatheoretic techniques developed by logicians.

However, they do not allow exceptions to inheritance, and exclude the possibility to represent concepts in prototypical terms. From this point of view, therefore, such formalisms can be seen as a revival of the classical theory of concepts. As far as prototypical information is concerned, such formalisms offer only two possibilities: representing it resorting to tricks or ad hoc solutions, or, alternatively, ignoring it. For obvious reasons, the first solution in unsuitable: it could have disastrous consequences for the soundness of the knowledge base and for the performances of the entire system. The second choice severely reduces the expressive power of the representation. In information retrieval terms, this could severely affect, for example, the system's recall. Let us suppose that you are interested in documents about flying animals. A document about birds is likely to interest you, because most birds are able to fly. However, flying is not a necessary condition to being a bird (there are many birds that are unable to fly). So, the fact that birds usually fly cannot be represented in a formalism that allows only the representation of concepts in classical terms, and the documents about birds will be ignored by your query.

Nowadays, DLs are widely adopted within many fields of application, in particular within the area of ontology representation. For example, OWL (Web Ontology Language)[1] is a formalism in this tradition, which has been endorsed by the World Wide Web Consortium for the development of the Semantic Web. However, DL formalisms leave unsolved the problems of representing concepts in prototypical terms.

Within the field of logic oriented KR, rigorous approaches exist, designed to make it possible the representation of exceptions, and that therefore are, at least in principle, suitable for dealing with (some aspects of) "non-classical" concepts. Examples are fuzzy and non-monotonic logics. Therefore, the adoption of logic oriented semantics is not necessarily incompatible with the representation of prototypical effects. Various fuzzy and non-monotonic extensions of DL formalisms have been proposed. Nevertheless, such approaches pose various theoretical and practical problems, which in part remain unsolved.

---

[1] http://www.w3.org/TR/owl-features/

Examples of the integration of fuzzy logic in DLs and in ontology oriented formalisms are Stoilos et al. (2005) and Bobillo and Straccia (2009). These approaches are surely valuable in many respects. However, from the standpoint of conceptual representation, it is well known (Osherson and Smith 1981) that fuzzy approaches to prototypical knowledge encounter severe difficulties with compositionality.

Various proposal of non-monotonic extensions of DLs have been advanced in the last two decades. For example, Baader and Hollunder (1995) extend the *ALCF* language with Reiter's defaults; Donini et al. (2002) propose an extension of DLs with two non monotonic epistemic operators; Bonatti et al. (2006) extend DLs with circumscription. However, seldom such extensions are rather counterintuitive, and often have bad computational properties. (For more references, and for a more detailed account, see Frixione and Lieto 2010.)

As a possible way out, we outline here a tentative proposal that goes in a different direction, and that is based on some suggestions coming from empirical cognitive science research. Within the field of cognitive psychology, different positions and theories on the nature of concepts are available. Usually, they are grouped in three main classes, namely: prototype views, exemplar views and theory-theories (see e.g. Murphy 2002; Machery 2009). All of them are assumed to account for (some aspects of) prototypical effects in conceptualisation.

According to the prototype view, knowledge about categories is stored in terms of prototypes, i.e. in terms of some representation of the "best" instances of the category. For example, the concept CAT should coincide with a representation of a prototypical cat. In the simpler versions of this approach, prototypes are represented as (possibly weighted) lists of features.

According to the exemplar view, a given category is mentally represented as set of specific exemplars explicitly stored within memory: the mental representation of the concept CAT is the set of the representations of (some of) the cats we encountered during our lifetime.

Theory-theory approaches adopt some form of holistic point of view about concepts. According to some versions of the theory-theories, concepts are analogous to theoretical terms in a scientific theory. For example, the concept CAT is individuated by the role it plays in our mental theory of zoology. In other version of the approach, concepts themselves are identified with micro-theories of some sort. For example, the concept CAT should be identified with a mentally represented micro-theory about cats.

These approaches turned out to be not mutually exclusive. Rather, they seem to succeed in explaining different classes of cognitive phenomena, and many researchers hold that all of them are needed to explain psychological data (see again Murphy 2002; Machery 2009). In this perspective, we propose to integrate some of them in computational representations of concepts. More precisely, we focus on prototypical and exemplar based approaches, and propose to combine them in a hybrid representation architecture in order to account for category representation and prototypical effects (for a similar, hybrid prototypical and exemplar based proposal in a different field, see Gagliardi 2008). We do not take into consideration here the theory-theory approach, since it is in some sense more vaguely defined if compared to the other two points of view. As a consequence, its computational treatment seems at present to be less feasible.

Prototype and exemplar based approaches to concept representation are, as mentioned above, not mutually exclusive, and they succeed in explaining different phenomena. Exemplar based representations can be useful in many situations. According to various experiments, it can happen that instances of a concept that are rather dissimilar from the prototype, but are very close to a known exemplar, are categorized quickly and with high confidence. For example, a penguin is rather dissimilar from the prototype of BIRD. However, if we already know an exemplar of penguin, and if we know that it is an instance of BIRD, it is easier for us to classify a new penguin as a BIRD. This is particularly relevant for concepts (such as FURNITURE, or VEHICLE) whose members differ significantly from one another.

Exemplar based representations are easier and faster to acquire, when compared to prototypes. In some situations, it can happen that there is not enough time to extract a prototype from the available information. Moreover, the exemplar based approach makes the acquisition of concepts that are not linearly separable easier (Medin and Schwanenflugel 1981). In the following section we shall review some of the available empirical evidence concerning prototype and exemplar based approaches to concept representation in psychology.

2 Exemplars vs. Prototypes in Cognitive Psychology

As anticipated in the previous section, according to the experimental evidence, exemplar models are in many cases more successful than prototypes. Consider the so-called "old-items advantage effect". It consists in the fact that already known items are usually more easily categorized than new items that are equally typical (see Smith and Minda 1998 for a review). For example: it is easier for me to classify my old pet Fido as a dog (even supposing that he is strongly atypical) than an unknown dog with the same degree of typicality. This effect is not predicted by prototype theories. Prototype approaches assume that people abstract a prototype from the stimuli presented during the learning phase, and categorize old as well as new stimuli by comparing them to it. What matters for categorization is the typicality degree of the items, not whether they are already known or not. By contrast, the old-item advantage is banal to explain in the terms of the exemplar paradigm.

This is correlated to a further kind of empirical evidence in favour of exemplar theories. It can happen that a less typical item can be categorized more quickly and more accurately than a more typical category member if it is similar to previously encountered exemplars of the category (Medin and Schaffer 1978). Consider the penguin example mentioned in the previous section: a penguin is a rather atypical bird. However, let us suppose that some exemplar of penguin is already stored in my memory as an instance of the concept BIRD. In this case, it can happen that I classify new penguins as birds more quickly and more confidently than less atypical birds (such as, say, toucans or hummingbirds) that I never encountered.

Another important source of evidence for the exemplar model stems from the study of linear separable categories (Medin and Schwanenflugel 1981). Two categories are linearly separable if and only if it is possible to determine to which of them an item belongs by summing the evidence concerning each attribute of this item. For example, let us suppose that two categories are characterized by two attributes, or dimensions, corresponding to the axes in fig. 1. These categories are linearly separable if and only if the category membership of each item can be determined by summing its value along the x and y axes, or, in other terms, if a line can be drawn, which separates the members of the categories.

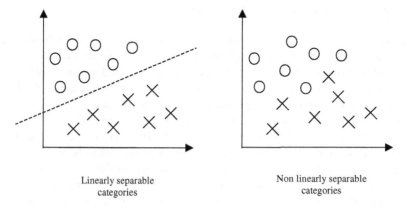

Linearly separable
categories

Non linearly separable
categories

Fig. 1

According to the prototype approach, people should find it more difficult to form a concept of a non-linearly separable category. Subjects should be faster at learning two categories that are linearly separable. However, Medin and Schwanenflugel (1981) experimentally proved that categories that are not linearly separable are not necessarily harder to learn. This is not a problem for exemplar based theories, which do not predict that subjects would be better at learning linearly separable categories. In the psychological literature, this result has been considered as a strong piece of evidence in favour of the exemplar models of concept learning.

The above mentioned results seem to favour exemplars against prototypes. However, other data do not confirm this conclusion. Moreover, it has been argued that many experiments favourable to the exemplar approach rest on a limited type of evidence, because in various experimental tasks a very similar category structure has been employed (Smith and Minda 2000). Nowadays, it is commonly accepted that that prototype and exemplars are not competing, mutually exclusive alternatives. In fact, these two hypotheses can collaborate in explaining different aspects of human conceptual abilities (see e.g. Murphy 2002; Machery 2009).

An empirical research supporting the hypothesis of a multiple mental representation of categories is (Malt 1989). This study was aimed to establish if people categorize and learn categories using exemplars or prototypes. The empirical data, consisting in behavioral measures such as categorization probability and reaction time, suggest that subjects use different strategies to categorize. Some use exemplars, a few rely on prototypes, and others appeal to both exemplars and prototypes. A protocol analysis of subjects' descriptions of the adopted categorization strategy confirms this interpretation[2]. Malt writes (1989, 546-547):

> Three said they used only general features of the category in classifying the new exemplars. Nine said they used only similarity to old exemplars, and eight said that they used a mixture of category features and similarity to old exemplars. If reports accurately reflect the strategies used, then the data are composed of responses involving several different decision processes.

These findings are consistent with other well known studies, such as (Smith et al. 1997) and (Smith and Minda 1998). Smith et al. (1997) found that the performances of half of the subjects of their experiments best fitted the prototype hypothesis, while the performances of the other half were best explained by an exemplar model. Therefore, it is plausible that people can learn at least two different types of representation for concepts, and that they can follow at least two different strategies of categorization. Smith and Minda (1998) replicated these findings and, additionally, found that during the learning, subjects' performances are best fitted by different models according to the features of the category (e.g., its dimensions) and the phase of the learning process, suggesting that when learning to categorize artificial stimuli, subjects can switch from a strategy involving prototypes to a strategy involving exemplars. They also found that the learning path is influenced by the properties of the learned categories. For example, categories with few, dissimilar members favour the use of exemplar-based categorization strategies. Thus, psychological evidence suggests that, in different cases, we employ different categorization mechanisms.

Summing up, prototype and exemplar approaches present significant differences, and have different merits. We conclude this section with a brief summary of such differences. First of all, exemplar-based models assume that the same representations are involved in such different tasks as identification (e.g., "this is the Tower Bridge") and categorization (Nosofsky 1986). This contrasts with prototype models, which assumes that these tasks involve different kinds of representations. Furthermore, prototype representations synthetically capture only some central, and cognitively relevant, aspects of a category, while models based on exemplars are more analytical, and represent *in toto* the available knowledge concerning the instances of a given category.

This is related to another aspect of divergence, which pertains the categorization process. Both prototype and exemplar models assume that the *similarity* between prototypical/exemplar representations and target representations is computed. The decision of whether the target belongs to some category depends on the

---

[2] A protocol analysis consists in recording what the subjects of an experiment say after the experiment about the way in which they performed the assigned tasks.

result of this comparison. However, important differences exist. According to the prototype view, the computation of similarity is usually assumed to be *linear*. Indeed, since prototypes are synthetic representations, all information stored in them is relevant. Therefore, if some property is shared by the target and by some prototype, this is sufficient to increase the similarity between them, independently from the fact that other properties are shared or not. On the contrary, an exemplar based representation includes information that is not relevant from this point of view (typically, information that idiosyncratically concerns specific individuals). As a consequence, the computation of similarity is assumed to be *non-linear*: an attribute that is shared by the target and by some exemplar is considered to be relevant only if other properties are also shared.

Prototypes and exemplar based approaches involve also different assumptions concerning processing and memory costs. According to the exemplar models, a category is mentally represented by storing in our long term memory many representations of category members; according to prototype theorists, only some parameters are stored, which summarize the features of a typical representative of the category. As a consequence, on the one hand, prototypes are synthetic representations that occupy a smaller memory space. On the other hand, the process of creating a prototype requires more time and computational effort if compared to the mere storage of knowledge about exemplars, which is computationally more parsimonious, since no abstraction is needed.

3 A Hybrid Prototype-Exemplar Architecture

Given the evidence presented in the above section, we make the hypothesis that a hybrid approach to concept representation that integrates prototypical and exemplar based approaches could enhance the performances of ontological knowledge bases. In this section we outline the proposal of a possible architecture for concept representation, which takes advantage of the suggestions presented in the sections above. It is based on a hybrid approach, and combines a component based on a DL with a further component that implements prototypical representations.

Concepts in the DL component are represented as in fig. 2. As usual, every concept can be subsumed by a certain number of superconcepts, and it can be characterised by means of a number of attributes, which relate it to other concepts in the knowledge base. Restrictions on the number of possible fillers can be associated to each attribute. Given a concept, its attributes and its concept/superconcept relations express necessary conditions for it. DL formalisms make it possible to specify which of these necessary conditions also count as sufficient conditions.

Since in this component only necessary/sufficient condition can be expressed, here concepts can be represented only in classical terms: no exceptions and no prototypical effects are allowed. Concepts can have any number of individual instances, that are represented as individual concepts in the taxonomy.

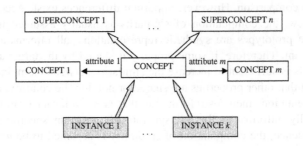

Fig. 2

As an example, consider the fragment of network shown in fig. 3.

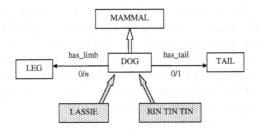

Fig. 3

Here the concept DOG is represented as a subconcept of MAMMAL. Since DL networks can express only necessary and/or sufficient conditions, some details of the representation are very loose. For example, according to fig. 3, a DOG may or may not have a tail (this is the expressed by the number restriction 0/1 imposed on the attribute *has_tail*), and has an unspecified number of limbs (since some dogs could have lost limbs, and teratological dogs could have more than four legs). LASSIE and RIN TIN TIN are represented as individual instances of DOG (of course, concepts describing individual instances can be further detailed, fully specifying for example the values of the attributes inherited from parent concepts).

Prototypes describing typical instances of concepts are represented as data structures that are external to the DL knowledge base. Such structures could, for example, be lists of (possibly weighted) attribute/value pairs that are linked to the corresponding concept. Some attributes of the list should correspond to attributes of the DL concept, which value can be further specified at this level. For example, the prototypical dog is described as having a tail, and exactly four legs. Other attributes of the prototype could have no counterpart in the corresponding DL concept.

As far as the exemplar-based component of the representations is concerned, exemplars are directly represented in the DL knowledge base as instances of concepts. (It may also happen that some information concerning exemplars is represented outside the DL component, in the form of Linked Data – for this notion see below, in this section. Typically, this could be the case of "non symbolic" information, such as images, sounds, etc.)

It must be noted that prototypical information about concepts (either stored in the form of prototypes or extracted from the representation of exemplars) extends the information coded within the DL formalism: the semantic network provides necessary and/or sufficient conditions for the application of concepts. As a consequence, such conditions hold for every instance of concepts, and cannot be violated by any specific exemplar. Therefore, what can be inferred on the basis of prototypical knowledge can extend, but can in no way conflict with what can be deduced from the DL based component.

According to our proposal, the categorisation of a new exemplar should follow the following steps:

a) Perform deductive reasoning on the DL knowledge base. Purely deductive, monotonic categorization requires sufficient conditions for concepts to be available. Since, in the majority of commonsense knowledge domains, sufficient conditions for defining concepts are scarce (or, when available, they cannot be used for many practical purposes), we can hypothesize that in most cases this step will give poor results. However, if something can be deductively categorized, then this inference is definitive (in the sense that it is not defeasible), and steps b) and c) can be eschewed.
b) Compare the item to be classified to the prototypes associated to the concepts, and evaluate their degree of similarity. If, for some concept, such a similarity value exceeds a given threshold, then the exemplar can be tentatively categorized as an instance of the corresponding concept.
c) Compare the item to be classified to the exemplars of the concepts stored in the knowledge base, and, again, evaluate their degree of similarity.

Of course, categorization performed on the basis of steps b) and c) is always defeasible. Moreover, conflicts between b) and c) can arise.

In recent years, one of the main application areas for concept representation has been the development of formal ontologies for the semantic web. In the field of web ontology languages, the developments proposed above could be achieved within the framework of the so-called Linked Data approach. In the semantic web research community, the Linked Data perspective is assuming a prominent position (Bizer, Heath and Berners-Lee 2009). One of the main objectives of this approach is the integration of different data representations (often stored in different data sources) within a unique representational framework. This makes it possible to enlarge the answer-space of a query through the realization of "semantic bridges" between different pieces of data (and, often, between different data sources). Such integration is made possible through specific constructs provided by Semantic Web languages, such as OWL, SKOS etc.

The implementation of both the exemplar and prototype points of view can take advantage of the Linked Data approach. Let us consider prototypes. Concepts can be represented as classes in a formal ontology, based on a classical, compositional DL system. Prototypes can be associated to such representations; they can be implemented using the Open Knowledge-Base Connectivity (OKBC) protocol[3]. The knowledge model of the OKBC protocol is supported by Protegé Frames, an ontology editor that makes it possible to build frame representations (the so called Frame Ontologies). Since it is possible to export the Frame Ontologies built with Protegé in the OWL language, the connection between these two types of representation can be made using the standard formalisms provided by the semantic web community in the linked data perspective (e.g. using the owl:sameAs construct).

In this way, according to our hypothesis, different types of categorization processes can follow different paths: monotonic categorization involves only the DL ontology, while typicality-based categorization, which involves exemplars and prototypes, could also take advantage of Linked Data structures that are external to the compositional ontology.

The possibility of performing forms of non-monotonic reasoning (namely, non-monotonic categorization of instances) only outside the compositional component of the representation system is one of the main features of our proposal. Among other things, this solution makes it possible to avoid consistency problems in the compositional part[4], introducing at the same time within the ontology (intended in a broad sense) the possibility to expand the allowed types of reasoning.

Hybrid non-monotonic categorization, based on both prototypes and exemplars, should take advantage of suggestions from the field of machine learning, where the prototype-exemplar dichotomy in concept representation has been investigated. Consider for example the *PEL-C algorithm*, where *PEL-C* stands for *Prototype-Exemplar Learning Classifier* (Gagliardi 2010). The *PEL-C* is a hybrid instance-based algorithm used for machine learning tasks, which accounts for typicality effects in categorization using both prototypes and exemplars. It is based on a learning phase as well as a test phase, and it can also be adopted for both a semi-automatic ontology population as well as updating processes.

Of course, the advantage of associating prototypical knowledge to concepts is not limited to categorization. Consider for example a task as *property checking*. Property checking consists of answering questions such as "does the class A have the property b?". Let us suppose that a user runs an informational query[5] on a knowledge base representing fruits, in order to know which kind of citrus fruit is yellow (i.e. (s)he asks the knowledge base the question: "does any citrus fruit have the property of being yellow?"). Intuitively, the expected answer that fits the information needs of the user is "lemon". However, in the DL knowledge base, does not exist any form of citrus fruit that has the property of being yellow as a defining condition. Being yellow is not a necessary condition for being a lemon and, therefore, this property cannot be associated to the class LEMON of the DL ontology. However, from a cognitive point of view, the property of being yellow is relevant

---

[3] http://www.ai.sri.com/~okbc/

[4] This was one of the main problems both in frame based systems as well as in hybrid knowledge representation approaches.

[5] According to the information retrieval literature, *informational queries* are different form *transactional* and *navigational* queries. In informational queries , the intention of the user is to obtain specific information concerning a given object (Jansen et al 2008).

to characterize the concept LEMON. According to our hybrid approach, this piece of knowledge can be represented in the prototypical information associated to LEMON (either in terms of the value of the attribute *colour* of the corresponding prototype, or stored in the knowledge concerning exemplars). In this way, it is possible to retrieve the desired information from the prototype and/or exemplar part of the representation. Thus, given a query such as:

SELECT the CONCEPT citrus WHERE {?CONCEPT citrus :has colour : YELLOW}

the result returned from the DL representation will be null, while the "correct" answer (i.e. correct with respect to the intention of the user) will be generated from the prototypical component of the representation.

4 Conclusions and Further Developments

In conclusion, our claim is that a hybrid prototype/exemplar based representation of non classical concepts could make ontological representation of commonsense concepts more flexible and realistic, and more convenient for many applications, including information retrieval. As far as further developments are concerned, we are now investigating the possibility of adopting conceptual spaces (Gärdenfors, 2000) as an adequate framework for representing both prototypes and exemplars in many different contexts.

References

Baader, F., Calvanese, D., McGuinness, D., Nardi, D., Patel-Schneider, P.: The Description Logic Handbook: Theory, Implementations and Applications, 2nd edn. Cambridge University Press (2010)

Baader, F., Hollunder, B.: Embedding defaults into terminological knowledge representation formalisms. J. Autom. Reasoning 14(1), 149–180 (1995)

Bizer, C., Heath, T., Berners-Lee, T.: Linked Data - The Story So Far. International Journal on Semantic Web and Information System 5(3), 1–22 (2009)

Bobillo, F., Straccia, U.: An OWL Ontology for Fuzzy OWL 2. In: Rauch, J., Raś, Z.W., Berka, P., Elomaa, T. (eds.) ISMIS 2009. LNCS, vol. 5722, pp. 151–160. Springer, Heidelberg (2009)

Bonatti, P.A., Lutz, C., Wolter, F.: Description logics with circumscription. In: Proc. KR 2006, pp. 400–410 (2006)

Brachman, R., Levesque, H. (eds.): Readings in Knowledge Representation. Morgan Kaufmann, Los Altos (1985)

Brachman, R., Schmolze, J.G.: An overview of the KL-ONE knowledge representation system. Cognitive Science 9, 171–216 (1985)

Donini, F.M., Nardi, D., Rosati, R.: Description logics of minimal knowl-edge and negation as failure. ACM Trans. Comput. Log. 3(2), 177–225 (2002)

Frixione, M., Lieto, A.: The Computational Representation of Concepts in Formal Ontologies: Some General Consideration. In: Proc. KEOD 2010, Valencia, Spain, pp. 396–403 (2010)

Frixione, M., Lieto, A.: Representing Concepts in Formal Ontologies: Compositionality vs. Typicality Effects, to appear in Logic and Logical Philosophy (in press)

Gagliardi, F.: A Prototype-Exemplars Hybrid Cognitive Model of "Phenomenon of Typicality" in Categorization: A Case Study in Biologi-cal Classification. In: Proc. 30th Annual Conf. of the Cognitive Science Society, Austin, TX, pp. 1176–1181 (2008)

Gagliardi, F.: Cognitive Models of Typicality in Categorization with In-stance-Based Machine Learning. In: Practices of Cognition. Recent Researches in Cognitive Science, pp. 115–130. University of Trento Press (2010)

Gärdenfors, P.: Conceptual Spaces: The Geometry of Thought. The MIT Press/Bradford Books, Cambridge, MA (2000)

Jansen, B.J., Booth, D.L., Spink, A.: Determining the informational, navigational, and transactional intent of Web queries. Information Processing and Management 44, 1251–1266 (2008)

Machery, E.: Doing without Concepts. Oxford University Press, Oxford (2009)

Malt, B.C.: An on-line investigation of prototype and exemplar strategies in classification. Journal of Experimental Psychology: Learning, Memory, and Cognition 15(4), 539–555 (1989)

Medin, D.L., Schaffer, M.M.: Context theory of classification learning. Psychological Review 85(3), 207–238 (1978)

Medin, D.L., Schwanenflugel, P.J.: Linear separability in classification learning. J. of Exp. Psyc.: Human Learning and Memory 7, 355–368 (1981)

Minsky, M.: A framework for representing knowledge. In: Winston, P. (ed.) The Psychology of Computer Vision. McGraw-Hill, New York (1975); Also in Brachman & Levesque (1985)

Murphy, G.L.: The Big Book of Concepts. The MIT Press, Cambridge (2002)

Nosofsky, R.M.: Attention, similarity, and the identification categorization relationship. Journal of Experimental Psychology: General 115, 39–57 (1986)

Osherson, D.N., Smith, E.E.: On the adequacy of prototype theory as a theory of concepts. Cognition 11, 237–262 (1981)

Quillian, M.R.: Semantic memory. In: Minsky, M. (ed.) Semantic Information Processing. The MIT Press, Cambridge (1968)

Rosch, E.: Cognitive representation of semantic categories. Journal of Experimental Psychology 104, 573–605 (1975)

Smith, J.D., Murray, M.J., Minda, J.P.: Straight talk about linear separabili-ty. Journal of Experimental Psychology: Learning, Memory, and Cognition 23, 659–668 (1997)

Smith, J.D., Minda, J.P.: Prototypes in the mist: The early epochs of cate-gory learning. Journal of Experimental Psychology: Learning, Memory, & Cognition 24, 1411–1436 (1998)

Smith, J.D., Minda, J.P.: Thirty categorization results in search of a model. Journal of Experimental Psychology: Learning, Memory, and Cognition 26, 3–27 (2000)

Stoilos, G., Stamou, G., Tzouvaras, V., Pan, J.Z., Horrocks, I.: Fuzzy OWL: Uncertainty and the Semantic Web. In: Proc. Workshop on OWL: Experience and Directions (OWLED 2005). CEUR Workshop Proceedings, vol. 188 (2005)

Wittgenstein, L.: Philosophische Untersuchungen. Blackwell, Oxford (1953)

From Logical Forms to SPARQL Query with GETARUNS

Rocco Tripodi and Rodolfo Delmonte

Abstract. We present a system for Question Answering which computes a prospective answer from Logical Forms produced by a full-fledged NLP for text understanding, and then maps the result onto schemata in SPARQL to be used for accessing the Semantic Web. As an intermediate step, and whenever there are complex concepts to be mapped, the system looks for a corresponding amalgam in YAGO classes. It is just by the internal structure of the Logical Form that we are able to produce a suitable and meaningful context for concept disambiguation. Logical Forms are the final output of a complex system for text understanding - GETARUNS - which can deal with different levels of syntactic and semantic ambiguity in the generation of a final structure, by accessing computational lexical equipped with sub-categorization frames and appropriate selectional restrictions applied to the attachment of complements and adjuncts. The system also produces pronominal binding and instantiates the implicit arguments, if needed, in order to complete the required Predicate Argument structure which is licensed by the semantic component.

1 Introduction

Nowdays, the need of the automatic processing of information on the web has become more and more relevant in order to develop applications able to cope with unstructured information.

Semantic Web (hence SW) is the project aiming at implementing a smarter web and has its fundament in a Tim Berners-Lee paper [1]. The article describes an Artificial Intelligence task applied to the web. The idea at the heart of the project is referencing things in the real world. The referencing procedure developed over the years is based on metadata and ontology. The metadata provides a computer-readable concept specification and the ontology provides a conceptual knowledge structure, which organizes concepts.

Rocco Tripodi · Rodolfo Delmonte
Ca' Bembo, Dorsoduro 1075 Università "Ca Foscari" 30123 – Venice, Italy
e-mail: {rocco,delmont}@unive.it

C. Lai et al. (Eds.): New Challenges in Distributed Inf. Filtering and Retrieval, SCI 439, pp. 183–196.
springerlink.com   © Springer-Verlag Berlin Heidelberg 2013

According to Wilchs [19] we could consider the SW to have an Information Extraction task at its heart. The SW task consists in relating entities to specific categories (e.g. Person, Place, Event, etc.). The formalism used to add facts in the SW is RDF (Resource Description Framework: http://www.w3.org/RDF/). We could, then, consider the SW datasets as *encyclopedia* (understood within the meaning given by Umberto Eco [11], as a network of interconnected cultural units), where we could find information about entities, and the reference could be considered as an attribution of meaning.

The W3C standard way to access a KB on the SW is SPARQL. SPARQL is used to express queries across data sources, whether the data is stored or viewed as RDF. In the Semantic Web, ontologies supply a machine-interpretable knowledge infrastructure. The real challenge does not only lie in constructing ontologies and keeping them up to date, but chiefly in linking them with the natural language [4]. In order to link automatically reference in text and entities in knowledge bases, a series of tools and heuristics are used for what can be called the semantic disambiguation task, i.e. discover the exact concept or the exact entities referenced in the text.

We present a system for Question Answering which computes a prospective answer from Logical Forms produced by a full-fledged NLP for text understanding, and then maps the result onto schemata in SPARQL to be used for accessing the Semantic Web.

This paper is divided into two parts. In the section below we focus on providing access to the SW through Natural Language. We discuss the problems we encountered and the solutions and strategies we adopt. The second part concerns Question Answering over Linked Data. We explain our question analysis approach and give examples of how our algorithm works.

1.1 Accessing the LOD Cloud through Natural Language: Problems

On the LOD Cloud [2] the information comes from different ontologies, lacking a semantic mapping among them, and many ontologies describe similar domains with different terminologies [10]. Such problems sketch two main points that we would like to address by means of semantic disambiguation technique and mapping process. Without NLP technique, access to the SW through Natural Language is allowed only using a short lexicon, which is made up of non homogeneous KBs labeling systems. This is due to the fact that a large KB has to handle with homonymy and synonymy problems. Liu [7] noted that DBpedia [3] contains a great number of disambiguation nodes. A disambiguation node is used to resolve conflicts when a single term can be the title of more than one article: for example the word "Mercury" can refer to several different things, including an element, a planet, an automobile brand, a record label, a NASA manned-spaceflight project, a plant, and a Roman god. Liu [7] explains that things linked by a disambiguation node are only related through rough homonymy. So when we look up a word in DBpedia we get a long list of possible candidates. Such problems are due to both word ambiguity and to the labeling system used. However, as Buitelaar claimed [5], the RDFS and OWL standards are not sufficient for the purpose of associating linguistic information with ontologies.

Besides the problem of homonymy, there is also the problem of synonymy. In DBpedia such problems are partially handled by the "redirect" property. A "redirect" property is a property (in the RDF formalism) that links a node A to a node B, where the node B is the preferred concept for A. That property is used in DBpedia to manage misspellings, alternative spellings, tenses, punctuation, capitalizations, etc. or to redirect sub-topic in broader context [15]. In that prospective we can see that the semantic content of the synonymy is not treated, and the access to the KB through natural language is limited to a short vocabulary. To avoid such a problem Buitelard et al., [5], have proposed a solution within the ontology markup standards. The idea behind this approach is to enter linguistic information inside the ontologies. Our approach, as will be explained below, does not attempt to modify the SW standards but tries to manage them by means of NLP and IE techniques.

By now we have discussed problems related to concept names, but a KB also contains names for classes and properties. The class names are common names which specify the collocation of a concept. The property links a concept with another concept, a class or a literal. As noted by Fu et al. [13] some relations in DBpedia have anomalous names that are hard to understand and therefore are difficult to use. Another problem concerns the fact that many relations share the same meaning, for example "dateOfBirth", "birthDate" and "datebirth" are three variant of the same concept. So when we want to retrieve all the entities with a particular property we have to collect all the various forms of the property. Similarly, DBpedia classes were extracted from different sources such as YAGO, UMBEL and Wikipedia. Only 170 were manually created for the project and are consistent with the DBpedia ontology [4]. Many extracted classes have the same problems of properties; besides, many classes express complex concepts with *n-ary* relations [5] such as:

1.CL   AncientGreekPhilosophers
2.CL   OlympicTennisPlayersOfTheUnitedStates

Classes of that kind have a complex semantics that is hard to use without a preprocessing phase. The first thing we do to handle these names is to split them into tokens. Then we proceed with GETARUNS [17, 18]. In particular, we analyzed them with a syntactic constituency parser and obtained the output below, where F3 is the label for fragments, SN stands for NounPhrase, SP for PrepositionalPhrase:

1.F3   f3-[sn-[Philosophers-n-sn, (mod)-[ancient_Greek-n-sn]]]
2.F3   f3-[sn-[olympic-ag-sn,tennis_players-n-sn,(mod)-[of-p-sp,the-art-sn,
       United_States-n-sn]]]

The analysis identifies the head and the modifiers of the head which is the governing name. At this point the heads must be disambiguated in order to be compared with the words in text. So we can use this information with contextual information. Modifiers are used to apply consistency checks.

Another step is done mapping the heads with synsets in WordNet, in order to expand the KB lexicon, for instance, the word "actress", in the question (1)*"Is Natalie Portman an actress?"* matches the class: dbpedia-owl:Actor, because "actress" share the same synset of "actor", as shown in the following term,

    dbp('Actor',[actor-n],[109765278,109767197]).

where we associated WordNet synset labels and DBPedia classes. In particular, DBP is a Prolog compound term, where the first element corresponds to a DBpedia label, the second element adds a POS tag to the label and the last element is a list with all synset labels. WordNet mapping allows us to use hyponymy relation, for instance, the word "wife" in the question (2)*"Who was the wife of President Lincoln?"* matches the property: dbpedia-owl:spouse, because there is an hyponymy relation between "wife" and "spouse".

1.2 Accessing the Semantic Web through Natural Language: Technique

Sowa [17] asserts that, each ontology, for practical application, must have a mapping, direct or indirect, related to and deriving from natural languages, because human knowledge is developed around human language. So an useful ontology must support a systematic mapping to and from natural languages, because such a bridge could break the static nature of a KB and make it flexible. The lack of this bridge has by now failed to achieve the hoped results in Artificial Intelligence and Knowledge Management [17].

What we have in mind is the assumption that an ontology reflects the background knowledge used in writing, reading and thinking [4]. In fact a text tells the reader which ontology to use to understand it [4]. The background knowledge, taken for granted by the author, is useful because can be used by a NLP application in order to decide a particular word sense. Word Sense Disambiguation (hence WSD) techniques use the notion of *context* in order to decide a particular word sense. A context could differ widely across WSD methods. One may consider a whole text, a word window, a sentence or some specific words [12]. Such techniques are necessary to access a static KB because the concepts are static objects; however knowledge can then be used and developed by reasoning. This approach comes from the *dynamic construal of meaning* (DCM) [7] approach, that we follow. The fundamental assumption of DCM is that the meaning of a word changes as it is used in different contexts or language games [17].

According to Chierchia [6] we consider the computation of meaning as a set of rules that determine the reference of words. We consider common names as classes, determiners as restrictions on classes, entities as referents and verbs as relations between entities and classes. This scheme is compatible with the RDF structure and can also serve as a bridge between natural language and KBs. Our approach is also related to Wittgenstein's language games [20], in that we assume we need to use patterns of words, to access an ontology. The RDF triples are atomic facts with a simple semantic. The meaning of each fact is the result of the meaning of three components:

- Classes: a class could be represented by a common name. When we talk about presidents, trees, cars, or carpenters, we are talking about classes of entities.
- Entities: we intend an entity as his reference. To access an entity we use his label and the disambiguation is done by one or more classes to which the entity belongs.
- Properties: are simple or complex relations between entities, classes and literal. We need to disambiguate a property and get contextual information from it.

With our approach, we want to extract information about the meaning of text. Particularly we want to understand what specific entities are mentioned in the text. To do this we use IE techniques to identify the named entities. We can use their names as labels to access a KB in order to get all the information regarding the entities. But as we noted above the same label could refer to several entities. The solution is to use contextual information. For instance, in the following example taken from the RTE5 challenge dataset:

(CNN) -- Malawians are rallying behind **Madonna** as she awaits a ruling Friday on whether she can adopt a girl from the southern African nation. **The pop star**, who has three children, adopted a son from Malawi in 2006. [...]

---

when we find an ambiguous entity (the pop start) we look for information that could disambiguate it. In this case, the singular definite expression "the pop star" is used to specify the entity Madonna. The definite expression consists of a determiner and a common noun that in our approach correspond to a class. At this point we have to establish which class could be associated with the noun found. This step corresponds to a WSD procedure, which serve as a bridge between natural language and KB. This approach is particularly useful in coreference resolution task where we have an identical name but different properties. In this way, coreference resolution is performed in parallel with entity identification. Consider another example below, with a text taken from the same RTE5 dataset:

[...] The volcano had erupted four times on Friday, billowing ash up to 51,000 feet up into the air. These are the latest in a series of <u>eruptions</u> from Mount Redoubt, [...]. The Alaska Volcano Observatory set its alert level at red, the highest possible level, meaning that an <u>eruption</u> is imminent, [...]

In this example the name "Mount Redoubt" could refer to different entities: Mount Redoubt (Alaska), Mount Redoubt (Washington), Redoubt Mountain (Canada), but the characteristic of being a volcano belongs only to one entity: :Mount_Redoubt. We use abduction to guess a new hypothesis that explains some fact. More on this in the following sections.

2 Question Answering over Linked Data

We start from the assumption that, any system for Information Extraction, or Question Answering, working under the hypothesis of open domain, unlimited vocabulary, and unstructured text, needs access to world knowledge. The encyclopedic knowledge we are referring to is the one that could be represented by web KB and in particular by the LOD project. Accessing KBs is done with the RDF triples structure in mind, which would correspond strictly to a predicate-argument structure; and the disambiguation task is done using background information derived from the text.

2.1 Question Analysis

As said above, question analysis is performed using GETARUNS [17, 18], the system for text understanding developed at the University of Venice, which is organized as a pipeline that includes two versions of the system: what we call the Partial and the Deep GETARUNS.

The system is based on LFG (Lexical Functional Grammar) theoretical framework and has a highly interconnected modular structure. The Closed Domain version of the system is a top-down depth-first DCG (Definite Clause Grammars) based parser written in Prolog Horn Clauses, which uses a strong deterministic policy by means of a lookahead mechanism with a WFST (Weighted Finite State Transducer) to help recovery when failure is unavoidable due to strong attachment ambiguity.

It is divided up into a pipeline of sequential but independent modules which realize the subdivision of a parsing scheme as proposed in LFG theory where a c-structure is built before the f-structure can be projected by unification into a DAG (Direct Acyclic Graph). In this sense we try to apply in a given sequence phrase-structure rules as they are ordered in the grammar: whenever a syntactic constituent is successfully built, it is checked for semantic consistency. In case the governing predicate expects obligatory arguments to be lexically realized they will be searched and checked for uniqueness and coherence as LFG grammaticality principles require.

Logical Forms derived from DAGs or f-structure sentence level representations are simplified in order to be useful for the question answering task. In particular, we come up with a non-recursive linear representation at propositional level where we introduce prefixes for each semantic head which are very close to DRS-conditions: PRED, QUANT, CARD, PLUR, ARG, MOD, ADJ, FOC. Where FOC contains the question type derived from a mapping of the Wh- word, its possible nominal or adjectival head and a restricted set of semantic general classes, like MEASURE, MANNER, QUANTITY, REASON etc.

2.2   *From Logical Form to SPARQL Query*

Our system produces a LF of natural language questions by means of GETARUNS. From LF, the system extracts the semantic elements needed to produce a SPARQL query that is then used to address LOD endpoint.

LFs produced by GETARUNS are all expressed as complex Prolog terms, and can be decomposed into three subparts: there is a Pred - the main verb predicate of the question -, a Focus - this is the question head expressed in the question which may correspond to an interrogative pronouns or may have a nominal head -, and then Arguments - this slot contains argument head and its internal modifiers and attributes like Quantifier, Cardinality, Plural. This slot may also contain other Arguments or entities and so on recursively. For instance, consider the following example: (3) *Which are the presidents of the United States of America?*

> Pred: [be], Focus: [person],
> Arg: [president/theme_bound-[['United_States_of_America']]]

As can be gather from the example, the Question is decomposed into three subelements, these are then used to build the SPARQL query. Predicate [be] can be regarded as the fact "belonging to a class". Focus [person] tells us that the reply foreseen by the question must be of Type Person, important feature which is easily expressed in SPARQL. We then look for the elements in Arg inside the two ontologies, DBPedia and YAGO [18] and we obtain the class: yago:PresidentOfTheUnitedStates. At this point, we can start building the query according to the schema, [?x a [Focus] . ?x a [Class]].

As explained above, there is no unique way of expressing the relation between properties and classes, and Person may belong to a number of different classes that have the same meaning. In order to cover the all of them in the KB we need to address them all in the query and consequently we come up with a multiple recursive query of the kind that we show below, where triples are conjoined by the clause UNION.

select distinct ?x ?string WHERE{
    { ?x a dbpedia-owl:Person . ?x a yago:PresidentsOfTheUnitedStates} union
    { ?x a foaf:person . ?x a yago:PresidentsOfTheUnitedStates} union
    { ?x a yago:Person100007846 . ?x a yago:PresidentsOfTheUnitedStates}
OPTIONAL { ?x rdfs:label ?string . FILTER (lang(?string) = "en")}}

In some cases, no useful class can be derived from Args produced by the LF. In that case, we need to introduce what can be regarded as FILTERS, which we derive from quantifiers and other restrictions to predicates, in order to narrow down the search, as for instance in the question: (4) *Who has been the 5th president of the United states of America?*

> Pred: [be], Focus: [person], Arg: [[president],card-'5th',['United_States']]

where we understand that the element individuated by Card, "5th", behaves like a restriction that operated on the class yago:PresidentsOfTheUnitedStates. Since there is no way to express such a restriction in SPARQL, we create a FILTER that looks into short literals for the specific word "5th", "president", "United States". This FILTER will be added to the previous query, like this:

?x ?prop ?lbl .
    FILTER (?prop != dbpedia-owl:abstract && ?prop != rdfs:comment && regex(?lbl, "(^|)president(|$)","i") && regex(?lbl, "(^|)5th(|$)","i") && regex(?lbl, "(^|)United States(|$)","i")).

YES/NO QUESTIONS

In case the LF does not produce a Focus element, the system understands that the question type is yes/no. In this case, the system will create a query of type ASK, which is meant to verify the existence of one or more RDF triples. Suppose the question is the following, (5) *Is Christian Bale starring in Batman Begins?*

> Pred: [be], Focus: [], Arg:['Christian_Bale'/theme_bound-[mod-[pred-[star], ['Begins'/theme-[mod-[pred-['Batman']]]]]]]

by analyzing the Arg element we realize that there are two entities and one property. In the organization of the final query, we proceed by looking for entities first: this we do because we find it important to verify the existence of a given concept before proceeding to submit the actual query containing it. In this preliminary phase, we search for the concepts related to the entities "Christian Bale" and "Batman Begins" in order to contextualize them. Then we also look for the predicate "star" in a special mapping we built where DBPedia properties are linked to WordNet verb synsets. When building this mapping, we found out that in many cases there was no possible correspondance between the information present in WordNet and the amalgamated labels of DBPedia. So we had to proceed manually.

The ASK query we produce for the above example is based on the simple scheme, [Ent Prop Obj], which produces the following query

ASK {
  {:Christian_Bale dbpedia-owl:starring :Batman_Begins.} Union
  { :Batman_Begins dbpedia-owl:starring :Christian_Bale . }}

as can be seen, we reverse the order of the two arguments of the predicate STAR, because we do not know whether it is being used in the active or the passive form. In other questions we proceed by disambiguating a property contained in the LF before proceeding to build the corresponding query. This is the case of the example below, (6) *Who was the wife of President Lincoln?*

> Pred: [be], Focus: [person]
> Arg: [wife/theme_bound-[['Lincoln'/theme-[(mod)-[pred-['President']]]]]]

Here, the system finds at first one property "being wife" (which is not expressed as a class but as a DBPedia property) and another element which consists of a label [Lincoln] and a class [President]. This latter property helps us to disambiguate the entity expressed by the question, because it contextualizes the reference, and it allows us to recover the actual intended entity, i.e. Abraham_Lincoln, by means of the procedure previously indicated. In this query, the scheme is the following one: [?x Prop Ent], and it allows us to build the following query:

select distinct ?x ?string WHERE {
  { ?x dbpedia-owl:spouse :Abraham_Lincoln .} Union
  { :Abraham_Lincoln dbpedia-owl:spouse ?x .}
  OPTIONAL { ?x rdfs:label ?string . FILTER (lang(?string) = "en")}}

Also in this case we use the reversed version of the query, which counts as the logically derivable statement "President Lincoln has a wife x".

FILTERS: GRADABLE ADJECTIVES AND QUANTIFIERS

There are other special cases of queries which require some filtering of the results, as shown in questions where the relevant property is expressed by a comparative or superlative adjective as in,

(7). *What is the highest mountain?*
7LF. [[be],focus-[mountain],[mountain/theme_bound-[(mod)-[pred-[highest]]]]]
(8). *Which mountains are higher than the Nanga Parbat?*
8LF. [[be],focus-[mountain],[higher/theme_bound-['Parbat'/theme_bound-[
      (mod)-[pred-['Nanga']]]]]]]

In both cases we have a superlative which is mapped through a specific filter: in (7) we have a scheme like:

   ?x a Class. ?x prop ?value. ORDER BY DESC(?value) LIMIT 1

which is transformed in the following query:

select distinct ?x ?string WHERE {
  {?x a dbpedia:Mountain. ?x dbpedia-owl:elevation ?value. } Union
  {?x a dbpedia:Mountain. ?x dbpedia2:elevationM ?value. }
  OPTIONAL { ?x rdfs:label ?string . FILTER (lang(?string) = "en")}}
  ORDER BY DESC(?value) LIMIT 1

In (8) the presence of a superlative induces a slightly different scheme:

   ?x a Class. ent prop ?valueE. ?x prop ?valueX. FILTER (?valueX > ?valueE) .

which is transformed in the following query:

select distinct ?x ?string WHERE {
  {?x a :Mountain. dbpedia:Nanga_Parbat dbpedia2:elevationM ?y1.
   ?x dbpedia2:elevationM ?y2.}
  {?x a :Mountain . dbpedia:Nanga_Parbat dbpedia-owl:elevation ?y1.
   ?x dbpedia- owl:elevation ?y2.} FILTER (?y2 > ?y1) .
  OPTIONAL { ?x rdfs:label ?string . FILTER (lang(?string) = "en")}}

In this case, at first we recover the class to which the prospective answers belongs, by means of DBPedia ontology, and then, after we have analyzed the superlative, we look for the properties it may be referred to and the kind of filter to use. Properties are recovered by means of our mapping onto DBPedia. As to the mapping of the two adjectives "higher" and "highest", they will be mapped both onto dbpedia2:elevationM and dbpedia-owl:elevation; because they are understood as belonging to the domain of :Place, which is the class right superior to :Mountain.

Information present in the Focus element allow us to build expectations and filters for a specific type of answer. In particular in case we have a question like: (8) *How many films did Leonardo DiCaprio star in?*

8LF [[do],focus-[quantity],pred-[star],(mod)-[pred-['Leonardo_DiCaprio']],[films]]

The Focus [quantity] requires us to count the number of results obtained from the query. Building the query then is done by using the remaining part of the question, where we have an entity [Leonardo_diCaprio], a predicate [star], and a class name [films]. Eventually we come up with the following scheme: [?x a Class. ?x prop Ent], just because the Focus is not a class, we can use the class found in the Arg to produce the final query:

```
select count(?x) WHERE {
  ?x a dbpedia-owl:Film
  {:Leonardo_DiCaprio dbpedia-owl:starring ?x.} union
  {?x dbpedia-owl:starring :Leonardo_DiCaprio.} union
  {?x dbpedia-owl:starring "Leonardo DiCaprio"@en.} }
```

Here again we reverse subject and object and we add a third entry which is referred to the label associated to the name of the entity. In fact, in many cases, DBPedia refers to an entity with one of its label rather than with referring to a URI. This fact is the reason why we lose sometimes points in the computation of recall, since literals may be missing when we impose a certain class to results of the search.

PRED NOT [BE]

When the predicate used in the question is not a copular verb, we come up with different schemes, as for instance in:

(9).   *Which books were written by Danielle Steel?*
9LF.   [[write],focus-[book],['Steel'/ [(mod)-[pred-['Danielle']]]]]
(10).   *Which actors were born in Germany?*
10LF   [[bear],focus-[actor],adj-[pred-['Germany']]]

The underlying scheme would be, [?x a [Focus]. ?x Pred [Arg]] from which we build two different queries: in the first case,

```
select distinct ?x ?string WHERE {
  ?x a dbpedia-owl:Book
  {:Danielle_Steel dbpedia-owl:author ?x.} union
  {?x dbpedia-owl:author :Danielle_Steel.} union
  {?x dbpedia-owl:author "Danielle Steel"@en.}
  OPTIONAL {?x rdfs:label ?string . FILTER (lang(?string) = "en")}}
```

in the second example,

```
select distinct ?x ?string WHERE {
  ?x a dbpedia-owl:Actor
  {?x dbpedia-owl:birthDate:Germany.} union
  {?x dbpedia-owl:birthPlace :Germany.} union
  {?x dbpprop :birthPlace :Germany.} union
  {?x dbpprop:birthDate :Germany.} union
  {?x dbpprop:birthDate :Germany.} union
  {?x dbpprop:placeOfBirth :Germany.}
  OPTIONAL {?x rdfs:label ?string . FILTER (lang(?string) = "en")}}
```

In the latter case, as in previous ones, we added recursively as many triples as there are properties linked to the Pred. Also note that in this case, subject and object are not reversed, and this is due to the nature of the complement which is computed as ADJunct or Oblique and not as Object or Xcomp(element) or open complement for predicative structures.

PROBLEMS

In our system the major problems we had have been with the ability to recover complex concepts, as for instance in the question: (11) *"Give me all female German chancellors!"* where we try to decompose the meaning into three different but intertwined queries (?x Female. ?x German. ?x Chancellor). But we don't get desired results and the reason is that DBPedia does not contain the male/female distinction. Probably there are amalgams which can express the complex concept to be a woman and be the head of the German government, but at the moment, our mapping strategy has not been able to find a class for the concept. On the contrary, it worked fine in the case of yago:PresidentsOfTheUnitedStates and in many others.

Other problems regard the use of literals in place of unique identifiers. For instance in the question: (12)*"In which programming language is GIMP written?"*

12LF   [[write],focus-[programming_language],['GIMP']

we use the scheme [?x a [Focus]. ?x Prop Ent]. But this is not correct since the reply for this question (C and GTK+) is not expressed with two unique references but with a literal, and literals cannot belong to any class. In this case the system does not receive a result and a second scheme is used, which consists in the elimination of the Focus and the reversal of subject and object: [Ent Prop ?x]. But also in this case we jump into a problem because we use as Prop [write] and this verb has a mapping which does not allow us to obtained the desired result: in fact, the property needed to obtain the correct result (C and GTK+) is dbpprop:programmingLanguage, and it is very difficult to derive from the Pred element [write].

3 Evaluation

We have tested our system on the training set made available by QALD-1[1] workshop organizers. The training set contains 50 question expressed in natural language to submit to DBPedia. We obtained correct answers (Precision and Recall = 1) to 23 questions over 50, with a final overall Precision and Recall equal to 0.46.

We looked into the mistakes and found out that:

a. in 14 cases, we did build up an efficient and adequate query;
b. in 5 cases we obtains partial results F-Measure ranging 0.40-0.80
b. in 4 cases we got a Precision ranging 0.80-0.98;
c. in 5 cases we got a Recall ranging 0.85-0.99.

In case a. we did build up a query with our schemas; we need to implement new ones. In case b. we obtained partial results and the Recall ranged between 0.4-0.8 indicates that we need to refine our filters. In case c. results are due to the presence of literals, which duplicate reference to the same entity with different names though: this could be avoided building up filters that eliminate multiple reference. In case d. we did not get some results. We assume that this is due to the fact that DBPedia allows to refer to the same entity or concept using different properties which however were not present in our mapping, thus preventing some elements not to be included in our results. We obtained 23 correct answers over the 50 DBpedia questions, with a global precision, recall and F-Measure = 0,46.

The execution time of a query depends largely on the execution time of the SPARQL query, in particular when it contains many restrictions it takes a long time to return the data. For the DBpedia dataset discussed in this work the execution time range from 1.4 seconds to 35.3 seconds, and 11.1 seconds on average per question.

4 Related Work

The analysis of the related work focuses on natural language query approaches for the Linked Data Web. In that field we have selected three QA systems: PowerAqua [16], FREyA (Feedback, Refinement and Extended VocabularY Aggregation) [9] and Treo [12], which share with our approach some similarities.

PowerAqua takes as input a natural language query, translates it into a set of logical queries, which are then answered by consulting and aggregating information derived from multiple heterogeneous semantic sources [9]. It is divided in three components: the Linguistic Component, the Relation Similarities Services and the Inference Engine. The first uses GATE [8] to obtain a set of syntactic annotations associated with the input query and to classify the query into a category. The Relation Similarity Service Component uses lexical resources and vocabulary of the ontology to map them to ontology-compliant semantic markup

---

[1] http://www.sc.cit-ec.uni-bielefeld.de/qald-1

or triples. The Inference Engine filters the candidate ontologies and selects the ones that present potential candidates.

Freya is an interactive Natural Language Interface for querying ontologies. It uses syntactic parsing in combination with the ontology-based lookup in order to interpret the question, and involves the user if necessary. FREyA workflow comprise for steps. The first is the Syntactic Parsing, which uses the Stanford Parser [14] in order to identify Potential Ontology Concepts (POCs). The second step is an Ontology-based Lookup that links the POCs with ontological concepts (OCs). The Consolidation algorithm is the third step; it consists in resolving ambiguous OCs in the question. In cases when the system fails to automatically generates the answer it prompt the user with a dialog. This dialogs involves the user to resolve identified ambiguities in the question.

Treo is a natural language query mechanism for Linked Data. It starts with the identification of the entities in the natural language query. After detection, key entities are sent to the entity search engine which resolves pivot entities in the Linked Data Web. A pivot entity is a URI which represents an entry point for the spreading activation search in the Linked Data Web. After the entities present in the user natural language query are determined, the query is analyzed in the query parsing module. This module tries to transform the structure of the terms present in the query in the subject, predicate, object structure of RDF. Starting from the pivot node, the algorithm navigates through neighboring nodes in the Linked Data Web computing the semantic relatedness between query terms and vocabulary terms in the node exploration process. The query answer is built through the node navigation process.

5 Conclusion

The problem cases are due to problems that our system has encountered for a lack of a strong mapping to many DBpedia properties. We have to understand the meaning that some properties have in DBpedia and then to move that information to the system, as we have already done automatically with classes and WordNet synset. Word is underway to improve on the mapping from SPARQL and with properties.

References

1. Berners-Lee, T., Hendler, J., Lassila, O.: The Semantic Web, Scientific American (2001)
2. Berners-Lee, T.: Linked Data - Design Issues (2006), http://www.w3.org/DesignIssues/LinkedData.html
3. Bizer, C., Lehmann, J., Kobilarov, G., Auer, S., Becker, C., Cyganiak, R., Hellmann, S.: DBpedia – A Crystallization Point for the Web of Data. Journal of Web Semantics (2009)

4. Brewster, C., Ciravegna, F., Wilks, Y.: Background and foreground knowledge in dynamic ontology construction. In: Proceedings of the Semantic Web Workshop, SIGIR, Toronto, Canada (2003)
5. Buitelaar, P., Cimiano, P., Haase, P., Sintek, M.: Towards linguistically grounded ontologies. In: Procs. of European Semantic Web Conference (2009)
6. Chierchia, G.: Le strutture del linguaggio. Semantica (The structures of language. Semantics) il Mulino, Bologna (1997)
7. Cruse, D.A.: Microsenses, default specificity and the semantics-pragmatics boundary. Axiomathes 1, 1–20 (2002)
8. Cunningham, H., Maynard, D., Bontcheva, K., Tablan, V.: GATE: A Framework and Graphical Development Environment for Robust NLP Tools and Applications. In: Proc. of the 40th Anniversary Meeting of the Association for Computational Linguistics (ACL 2002), Philadelphia (2002)
9. Damljanovic, D., Agatonovic, M., Cunningham, H.: Natural Language Interfaces to Ontologies: Combining Syntactic Analysis and Ontology-Based Lookup through the User Interaction. In: Aroyo, L., Antoniou, G., Hyvönen, E., ten Teije, A., Stuckenschmidt, H., Cabral, L., Tudorache, T. (eds.) ESWC 2010. LNCS, vol. 6088, pp. 106–120. Springer, Heidelberg (2010)
10. Doan, A., Madhavan, J., Dhamankar, R., Domingos, P., Halevy, A.: Learning to Match Ontologies on the Semantic Web. VLDB Journal 12(4), 303–319 (2003)
11. Eco, U.: Dall'albero al labirinto. Studi storici sul segno e l'interpretazione (From the tree to the labyrinth), Bompiani (2007)
12. Freitas, A., Oliveira, J., Curry, E., O'Riain, S., Pereira da Silva, J.: Treo: Combining Entity-Search, Spreading Activation and Semantic Relatedness for Querying Linked Data. In: Proceedings of the 1st Workshop on Question Answering Over Linked Data (QALD-1) (May 30, 2011); Heraklion, Greece Co-located with the 8th Extended Semantic Web Conference
13. Fu, L., Wang, H., Yu, Y.: Towards Better Understanding and Utilizing Relations in DBpedia. International Journal of Web Intelligence and Agent Systems
14. Klein, D., Manning, C.: Fast Exact Inference with a Factored Model for Natural Language Parsing. In: Becker, S., Thrun, S., Obermayer, K. (eds.) Neural Information Processing Systems 2002. Advances in Neural Information Processing Systems, vol. 15, MIT Press (2003)
15. Liu, O.: Relation Discovery on the DBpedia Semantic Web. Framework (2009)
16. Lopez, V., Motta, E., Uren, V.S.: PowerAqua: Fishing the Semantic Web. In: Sure, Y., Domingue, J. (eds.) ESWC 2006. LNCS, vol. 4011, pp. 393–410. Springer, Heidelberg (2006)
17. Sowa, J.F.: The role of logic and ontology in language and reasoning. In: Poli, R., Seibt, J. (eds.) Theory and Applications of Ontology: Philosophical Perspectives, ch. 11, pp. 231–263. Springer, Berlin (2010)
18. Suchanek, F.M., Kasneci, G., Weikum, G.: Yago - A Core of Semantic Knowledge. In: 16th International World Wide Web Conference (2007)
19. Wilks, Y.: Information extraction as a core language technology. In: Pazienza, M.-T. (ed.) Information Extraction. Springer, Berlin (1997)
20. Wittgenstein, L.: Philosophical Investigations. Basil Blackwell, Oxford (1953)

A DHT-Based Multi-Agent System for Semantic Information Sharing

Agostino Poggi and Michele Tomaiuolo

Abstract. This paper presents AOIS, a multi-agent system that supports the sharing of information among a dynamic community of users connected through the Internet thanks to the use of a well-known DHT-based peer-to-peer platform: BitTorrent. In respect to Web search engines, this system enhances the search through domain ontologies, avoids the burden of publishing the information on the Web and guaranties a controlled and dynamic access to the information. The use of agent technologies has made the realization of three of the main features of the system straightforward: i) filtering of information coming from different users, on the basis of the previous experience of the local user, ii) pushing of some new information that can be of interest for a user, and iii) delegation of access capabilities, on the basis of a reputation network, built by the agents of the system on the community of its users. The use of BitTorrent will allow us to offer the AOIS systems to the hundreds of millions of users that already share documents through the BitTorrent platform.

1 Introduction

Nowadays, the Web is the most powerful means for getting information about any kind of topic. However, the Web assigns a passive role to the large part of its users. Therefore, when Internet must be used to allow the active sharing of information among the members of a community, the use of a peer-to-peer solution may provide several advantages [22].

This paper presents a system, called AOIS (Agents and Ontology based Information Sharing), trying to couple the features of peer-to-peer and multi-agent systems. The next section introduces related work on multi-agent systems for

Agostino Poggi · Michele Tomaiuolo
Dipartimento di Ingegneria dell'Informazione,
Università degli Studi di Parma
Viale U. P. Usberti 181/A, 43100 Parma, Italy
e-mail: {agostino.poggi,michele.tomaiuolo}@unipr.it

C. Lai et al. (Eds.): New Challenges in Distributed Inf. Filtering and Retrieval, SCI 439, pp. 197–213.
springerlink.com   © Springer-Verlag Berlin Heidelberg 2013

information retrieval. Section three describes the main features and the behaviour of the AOIS system. Section four describes how this system has been designed and implemented by using some well-known technologies and software tools. Section five briefly discusses the testing of the system. Finally, section six reports some concluding remarks, gives a short introduction about the first experimentation of the system and presents our future research directions.

2 Related Work

Multi-agent systems have always been considered one of the most important ingredients for the development of distributed information management systems and for proving the different services needed in such systems [13]. In particular, several interesting works demonstrate: how multi-agent systems are a suitable means for the management of information in a community of users, how they can take advantage of a peer-to-peer network for performing a distributed search of information and how the use of ontologies and user profile allows an improvement of the quality of their work.

DIAMS is a multi-agent system that provides services for users to access, manage, share and learn information collaboratively on the Web [5]. DIAMS can be considered one of the most complete multi-agent infrastructures for the management and retrieval of information in a community of users. In fact, it supports the searching and retrieval of the information from local and/or remote repositories and it encourages the collaboration among its users by supporting the sharing and exchange of information among them.

ACP2P (Agent Community based Peer-to-Peer) is an information retrieval system that uses agent communities to manage and search information of interest to users [18]. In the ACP2P system, an agent works as a delegate of its user and searches for information that the user wants by coupling the typical propagation of the query on the peer-to-peer infrastructure. It supports the community with the identification of the agents that may have such information through the use of the experience gained in its previous interactions. The experimental results of the use of the ACP2P system demonstrated that the use of the agent experience provides a higher accuracy in retrieving information.

CinemaScreen is a recommender system, which combines collaborative filtering and content-based filtering [26]. The first method requires matching a user with other users with similar behaviours and interests. The second method requires matching the items on the basis of their characteristics (CinemaScreen, in particular, deals with genres, actors, directors etc.). While both mechanisms exhibit weaknesses in particular situations, their combination allows better performances since the very beginning of the system activity. The system is built in the form of an intelligent agent, but apparently it is modelled as an essentially centralized system.

On the other hand, pSearch is a decentralized information retrieval system [30]. In this system, which is P2P but non-flooding, document indices are distributed through the network according to a classification of document content. The document semantics is generated and managed through a technique called Latent

Semantic Indexing [34]. The resulting system is proven to be efficient in the number of nodes to contact to perform a search.

In [11] a social resource sharing system is presented. In this case, it uses a form of lightweight knowledge representation, called folksonomy. In fact, the conceptual structures of 'taxonomy' are created bottom-up by 'folks', thus creating an emergent semantics, instead of using the more rigid approach of the traditional Semantic Web

Sanchez and his colleague proposed an integrated agent-based ontology-driven multi-agent system that automatically retrieves Web pages that contain data relevant to the main concepts of a specific domain [27]. The multi-agent system is based on the use of a Web-based ontology learning method able to automatically build ontologies for any domain [20], and then on a set of agents that use such ontologies for the retrieval, filtering and classification of information.

3   AOIS

AOIS (Agents and Ontology based Information Sharing) is a multi-agent system composed of different agent platforms connected through the internet that supports the sharing of information among a community of users. Each agent platform acts as a "peer" of the system and is based on five agents: a personal assistant (PA), a repository manager (RM), an information finder (IF), an information pusher (IP), and a directory facilitator (DF). Figure 1 shows the AOIS system architecture.

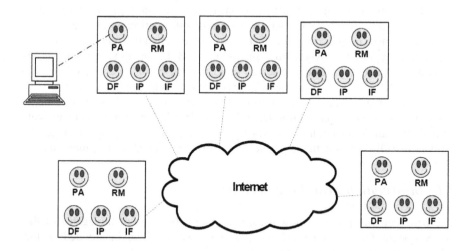

Fig. 1 AOIS system architecture.

A personal assistant (PA) is an agent that allows the interaction between the AOIS system and the user. This agent receives the user's queries, forwards them to the available information finders and presents the results to the user. Moreover, a PA allows the user to be informed about the new information that other users made available and that may be of her/his interest. Finally, a PA maintains the information that a user may share allowing her/him to add and remove information in a repository on the basis of the topics of interest of the user.

A repository manager (RM) is an agent that builds and maintains both the indexes for searching information and the ontologies describing the topics of interest of its user. Each time the user adds or removes some information, the RM updates the corresponding index and ontology.

An information finder (IF) is an agent that searches information on the repository contained into the computer where it lives and provides this information both to its user and to other users of the AOIS system. An IF receives users' queries, finds appropriate results, on the basis of both the queries and the topic ontology, and filters them on the basis of its user's policies (e.g., the results from non-public folders are not sent to other users).

An information pusher (IP) is an agent that monitors the changes in the local repository and pushes the new information to the PA of the users whose previous queries match such information.

Finally, the directory facilitator (DF) is responsible to register the agent platform in the AOIS network. The DF is also responsible to inform the agents of its platform about the address of the agents that live in the other platforms available on the AOIS network (e.g., a PA can ask about the address of the active IF agents).

The exchange of information among the users of an AOIS system is driven by the creation of both a search index and an ontology for each topic. The search index allows the ranking of information on the basis of the terms contained in a query. The ontology allows to identify additional information on the basis of the terms contained in the ontology that have some semantic relationships (i.e., synonyms, hyponyms, hypernyms, meronyms and holonyms) with the terms contained in the query. Both the search index and the ontology are automatically built by the RM on the basis of the information stored in the topic repository.

The following subsections describe the behaviour of the AOIS system through five practical scenarios and introduce a detailed description of how members can be added to an AOIS community and how security and privacy are managed to show how AOIS copes with the problems of working in a real open community.

3.1 Information Searching Scenario

The first scenario describes how a user can take advantage of the agents of the AOIS system for searching information. This scenario can be divided in the following five phases (see also figure 2):

1) a user requests her/his PA to search information on the basis of a topic, a set of keywords. The PA asks the DF for the addresses of available IF agents and sends the topic and the keywords to such agents (information search request phase);

2) each IF checks if the querying user has the access to at least a part of the information about the topic stored in the corresponding topic repository, and, if it happens, searches the information on the basis of both the received query and a set of additional queries obtained by replacing each keyword of the received query with the possible substitutes contained in the topic ontology. Moreover, the IF sends the received query to the local IP and RM agents: the IP adds the query to the profile of the corresponding remote user and the RM add the query keywords to the list of the keywords for updating the repository ontology (information search execution phase),

3) each IF filters the searching results on the basis of the querying user access permissions and sends the filtered list of results to the querying PA (information filtering and sending phase);

4) the querying PA orders the various results as soon as it receives them, omitting duplicate results and presents them to its user (information presentation phase);

5) after the examination of the results list, the user can ask her/his PA for retrieving the information corresponding to an element of the list. Therefore, the PA forwards the request to the appropriate IF, waits for its answer and presents the information to the user (information retrieval phase).

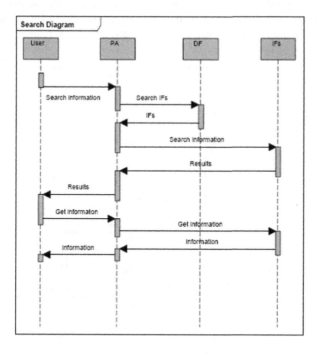

Fig. 2 Searching scenario UML sequence diagram.

3.2 Information Pushing Scenario

The second scenario illustrates how a user can take advantage of the AOIS system to be aware about the availability of new information of her/his interest. This scenario can be divided in the following five phases (see also figure 3):

1) a user requests her/his PA to add some information in a specific topic repository and the PA propagates the request to the RM (information addition request phase);
2) the RM adds the information in the repository, updates the indexes for the searching of information and then informs the IP about the new information (information addition phase);
3) the IP checks if such new information satisfy some queries maintained in the profiles of the remote users and, when happens, then the IP either sends such information to the PA of the remote user (if the corresponding AOIS platforms are alive), or maintains such an information until such a platform becomes alive again (information pushing phase);
4) of course, when a PA receives a list of pushing results, it presents them to its user (information presentation phase);
5) after the examination of the results list, the user can ask her/his PA for retrieving the information corresponding to an element of the list. Therefore, the PA forwards the request to the appropriate IF, waits for its answer and presents the information to the user (information retrieval phase).

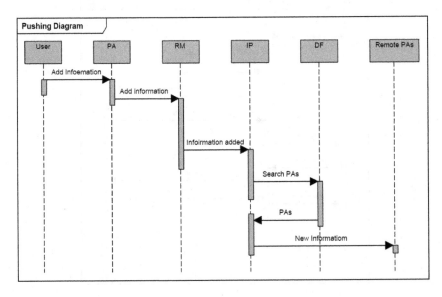

Fig. 3 Pushing scenario UML sequence diagram.

3.3 Repository Creation Scenario

The third scenario illustrates how a user can take advantage of the AOIS system for the creation of a repository for maintaining the information about a specific topic. This scenario can be divided in the following four phases:

1) a user requests her/his PA to create a repository for a specific topic indicating the set of terms (named ontology top terms) that describe such a topic and listing a set of information to store in the repository. The PA propagates the request to the RM (repository creation request phase);

2) the RM creates the repository, builds the topic ontology finding the semantic relationships (i.e., synonyms, hyponyms, hypernyms, meronyms and holonyms) among the top terms, adds the set of information, builds the search indexes and then informs the PA about the creation (repository creation phase);

3) the PA asks its user if she/he wants to populate the ontology with terms extracted from the information stored in the repository and the maximum permitted semantic distance between a new and a top term (ontology population request phase);

4) If the user enables the operation, the PA asks the RM to analyse the repository search indexes for finding the terms that are in direct or indirect relations with the top terms of the ontology. Of course, in the case of indirect relationship, each new term is only added if it satisfies the maximum semantic distance constraint (ontology population phase).

3.4 Repository Updating Scenarios

The forth and fifth scenarios illustrate how a user can take advantage of the AOIS system for updating both the search indexes of a repository and the related ontology.

The forth scenario is driven by the user that wants to add some information to a repository. This scenario can be divided in the following four phases:

1) the user requests her/his PA to add some information to a repository and the PA propagates the request to the RM (information addition request phase);

2) the RM adds the information in the repository, updates the indexes for the searching of information and then informs the PA about the new information (information addition phase);

3) the PA asks its user if she/he wants to populate the ontology with terms extracted from the new information added in the repository and the maximum permitted semantic distance between a new and a top term (ontology updating request phase);

4) If the user enables the operation, the PA asks the RM to analyse the repository search indexes for finding the terms that are in direct or indirect relations with the top terms of the ontology. Of course, in the case of indirect relationship, each new term is only added if it satisfies the maximum semantic distance constraint (ontology population phase).

The fifth scenario starts when the user logs to the system and her/his RM has some new keywords coming from last remote users queries. This scenario can be divided in the following two phases:

1) when the user logs the system, the PA gets the new keywords from the RM and asks its user if she/he likes to add some of them as top terms of the ontology (keywords selection request phase);

2) if the user selected some of the keywords to populate the ontology, the PA asks the RM to update the ontology finding the semantic relationships (i.e., synonyms, hyponyms, hypernyms, meronyms and holonyms) among the new and the old top terms (ontology updating phase);

3) Then the PA asks her/his user if she/he wants to populate the ontology with terms extracted from the information stored in the repository and the maximum permitted semantic distance between a new and a top term (ontology population request phase);

4) If the user enables the operation, the PA asks the RM to analyse the repository search index for finding the terms that are in direct or indirect relations with such new top terms of the ontology. Of course, in the case of indirect relationship, each term is only added if it satisfies the maximum semantic distance constraint (ontology population phase).

3.5   Community Management Scenario

The fifth scenario illustrates how an AOIS user can connect to an existing community and how a community can deal with new join requests.

This scenario can be divided in the following four phases:

1) the user has to be introduced into the community by a member who plays the role of introducer;

2) the new member is acknowledged by the introducer by receiving a proper token, which testifies the acceptance into the community;

3) once the new member has been acknowledged by the introducer, the latter also sends a list of all other members of the community, with their basic profile and contact information, to the new member;

4) the new member registers all members into the local list of contacts, and updates the information of the DF of her/his AOIS platform on the basis of the profiles received by the other members of the community;

5) the new member then sends a join request to all other members, together with all relevant credentials, including the token received from the introducer;

6) the other members of the community add the new user's profile to the local list of contacts and adds his services into the local DF.

3.6   Security and Privacy Management

The information stored into the different repositories of a AOIS network is not accessible to all the users of the system in the same way. In fact, it's important to avoid the access to private documents and personal files, but also to files reserved to a restricted group of users (e.g.: the participants of a project). The AOIS system takes care of users' privacy allowing the access to the information on the basis of the identity, the roles and the attributes of the querying user, as defined into a local knowledge base of trusted users. In this case, it is the user that defines who and in which way can access to her/his information. Moreover, the user can also allow grant the access to unknown users by enabling a certificate based delegation, built on a network of the users registered into the AOIS community. In this sense, the system completely adheres to the principles of trust management. For instance, if the user U_i enables the delegation and grants to the user U_j the access to its repository with capabilities C_0, and U_j in turn grants to the user U_k the access to its the repository with the same capabilities C_0 to the user U_k, then U_k can access U_i's repository with the same capabilities of U_j.

The definition of roles and attributes is made in a local namespace, and the whole system is, in this regard, completely distributed. Local names are distinguished by prefixing them with the principal defining them, i.e. a hash of the public key associated with the local runtime. Links among different local namespace, again, can be explicitly defined by issuing appropriate certificates. In this sense, local names are the distributed counterpart of roles in role based access control frameworks [14]. This model is centred on a set of roles. Each role can be granted a set of permissions, and each user can be assigned to one or more roles. A many to many relationship binds principals and the roles they're assigned to. In the same way, a many to many relationship binds permissions and the roles they're granted to, thus creating a level of indirection between a principal and his access rights. Like roles, local names can be used as a level of indirection between principals and permissions. Both local names and roles represent at the same time a set of principals and a set of permissions granted to those principals. But, while roles are usually defined in a centralized fashion by a system administrator, local names, instead, are fully decentralized. This way, they better scale to Internet-wide, peer-to-peer applications, without loosening in any way the principles of trust management.

In AOIS, the user can not only provide the permission to access his own files, but can also assign the permission to upload a new version of one or more existing files. In this case the PA informs his/her user about the updated files the first time he/she logs in. This functionality, together with the trust delegation, can be useful for the members of a workgroup involved in common projects or activities

4 Implementation

The AOIS system has been designed and implemented taking advantage of agent, peer-to-peer, information retrieval and security management technologies and, in particular, of five main components: JADE [3], BitTorrent DHT [6], Nutch [1], WordNet [17] and JAWS [28].

AOIS agent platforms have been realized by using JADE [3,4,31]. JADE is probably the most known agent development environment enabling the integration of agents and both knowledge and Internet-oriented technologies. Currently, JADE is considered the reference implementation of the FIPA (Foundation for

Intelligent Physical Agents) specifications [8]. In fact, it is available under an LPGL open source license, it has a large user group, involving more than two thousands active members, it has been used to realize real systems in different application sectors, and its development is guided by a governing board involving some important industrial companies.

The JADE development environment does not provide any support for the realization of real peer-to-peer systems because it only provides the possibility of federating different platforms through a hierarchical organization of the platform directory facilitators on the basis of a priori knowledge of the agent platforms addresses. Therefore, we extended the JADE directory facilitator to realize real peer-to-peer agent platforms networks thanks to DHT indexing mechanisms and popular file-sharing protocols.

In the first prototypes [23], we used JXTA protocols to augment JADE with the desired peer-to-peer features [9]. In fact, FIPA had acknowledged the importance of the JXTA protocols, and it had released some draft specifications for the interoperability of FIPA platforms connected to peer-to-peer networks. In particular, in the "FIPA JXTA Discovery Middleware Specification" a Generic Discovery Service (GDS) is described, to discover agents and services deployed on FIPA platforms working together in a peer-to-peer network. However, no advancement has then been made for these specifications and JXTA itself has not gained the expected popularity and maturity.

For these reasons, we turned to BitTorrent as one of the most widespread and solid file-sharing platform, which can be configured and extended to work in a completely decentralized fashion [6]. Actually, BitTorrent names both a file-sharing protocol and a particular application, implementing the protocol itself. Other applications, available for many existing hardware/software platforms, implement the protocol.

Basically, BitTorrent requires a tracker server to host so-called torrent files. A torrent file contains the updated list of seeds from which a particular resource can be obtained. For our purposes, i.e. to build a decentralized collaborative network, we preferred avoiding this basic approach. Moreover, in the recent past it has been proven vulnerable to both technical disruptions and legal actions. In fact, today some alternatives allow the realization of trackerless systems.

Azureus was the first application to introduce a Distributed Hash Table to supplement the centralized index. Today, this indexing mechanism is supported through a standard plug-in and is called Distributed Database (DDB). Vuze is an evolution of Azureus that uses such an approach [32]. The Vuze DDB is based on the Kademlia algorithms by Maymounkov and Mazières, which are essentially used to associate the hashes of files and chunks with their current locations (seeds), in a fully distributed fashion [16]. A widespread standard to share a reference to a file is a magnet-uri, which contains the hash of the file. Both nodes and shared files have globally unique, 160 bit long, identifiers. Each node maintains a small routing table with contact information for a small number of other nodes; the routing table is more detailed for closer nodes. The distance is measured according to the XOR metric defined by Maymounkov and Mazières. The information regarding the peers sharing a given file is stored on nodes with ID close to the hash of the file. When a node wants to download a file for which it knows the hash, it asks further information to the nodes it knows with ID closer to the file hash. These nodes answer with the list of peers downloading the file if they have such an information, otherwise they return a list of nodes with IDs even closer to the file hash, which should be queried afterwards.

Apart from basic file sharing protocols, however, a generic service advertisement system needs some mechanisms to discover a, possibly semantically enriched, service description, starting from some requested features and desired quality of service. In principle, keywords and tags can be associated with any file, and in particular with a service description, quite easily even over DHTs. In fact, some decentralized file sharing platforms use their DHT to implement two different indexes: one for associating seeds to file IDs, the other for associating file IDs with some keywords. However, the keyword index is hardly verifiable in an automated way and in the real world it proved to be particularly weak with respect to pollution and index poisoning attacks. Montassier et al. provide a measure of the credibility of the keyword index of the popular KAD network and show that around 2/3 of the contents are polluted [19]. For this reason, in current implementation of our service discovery system, keyword indexing is based on a DHT, but the keyword-service association is only trusted if provided by a trusted node, participating in the same collaborative network, and possibly other sources suggested by those trusted nodes. Specifically, each node can associate some attributes with the descriptions of the services it provides. This information is then published in the DHT under a unique key, which is obtained by combining the attribute and the node's identity in the system. Other than the attributes associated with a service in the DHT, a node can then analyze the obtained descriptions, in detail, to choose a particular service among those with the basic set of attributes.

Currently, also other applications support some form of DHT indexing. In particular, the BitTorrent application introduced a mechanism named Mainline DHT, which is also based on Kademlia. The queries available under the Mainline DHT allow a robust exchange of information and well support the BitTorrent file-sharing protocol. However, they are unsuitable for our purposes. In order for the DHT to map an attribute to some service descriptions, it is necessary to use arbitrary keys (the node identifiers combined with the attribute) on the DHT. Essentially, a couple of put/get queries would be needed, which would simply associate a given key to a given value on the DHT. Those queries, instead, are readily available under the Vuze DDB, and thus make it a preferable choice when implementing a generic service advertisement and discovery system. As a consequence, we decided to use Vuze DDB for both our logical DHTs, although in principle the DHT mapping hashes to the files that generated them could have been the Mainline DHT.

Regarding the Vuze platform, it has a modular architecture, where functionality can be added with plug-ins. The main application exposes to the plug-ins only a restricted programming interface, which is nonetheless sufficient for our purposes. Consequently, we decided to implement the service discovery system as a Vuze plug-in. As Vuze is implemented as a modular system, it is possible to run it without any graphical user interface, and thus to use its backend features inside other applications, too. In our case, the Vuze backend is used for realizing the needed discovery, location and sharing services inside a full agent-based system based on JADE. The multi-agent system acts as a Vuze plugin, while Vuze APIs are used as low level primitives for implementing the needed services inside the multi-agent system.

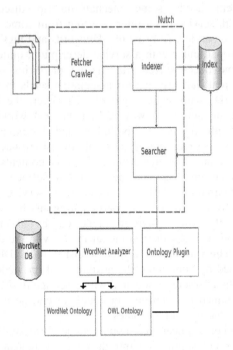

Fig. 4 Indexing and ontology management subsystem.

Even if there are some specific tools and software libraries for searching information in a local repository (see, for example, Beagle [2] and Google Desktop Search [10]), we adapted Nutch [1], an open source web-search software, for searching the local repository. It has been done because it is very easy to develop Nutch plugins for extending its capabilities (we used this feature for using its term extraction module for building the topic ontologies) and because is available a Nutch plugin, that extends keywords based search through the use of OWL ontologies [33]. Figure 4 shows a graphical description of the work done by the Nutch core software and by its two plugins for indexing and building the topic ontologies and for using them for searching information.

As introduced above, topic ontologies are built by a Nutch plugin. This plugin receives the terms extracted from the information to be indexed by the Nutch software. Then, accessing the WordNet lexical database [17,24] though the use of the JAWS Java software library [28], for each term it identifies the top terms of the ontology and the other terms extracted from the information that have some semantic relationships (i.e., synonyms, hyponyms, hypernyms, meronyms and holonyms). At the end of this process, all the terms that have a semantic distance greater than the one fixed by the user are removed and then the WordNet ontology is saved as an OWL file.

As introduces before, authentication and authorization are performed on the basis of the local knowledge base of trusted users, though they can be delegated to external entities through an explicit, certificate based delegation. In this sense, the system completely adheres to the principles of trust management. The definition of roles and attributes is also made in a local namespace, and the whole system is, in this regard, completely distributed. Local names are distinguished by prefixing them with the principal defining them, i.e., a hash of the public key associated with the local agent platform. Links among different local namespace, again, can be explicitly defined by issuing appropriate certificates. The theory of AOIS delegation certificates is founded on SPKI/SDSI specifications [7], though the certificate encoding is different. As in SPKI, principals are identified by their public keys, or by a cryptographic hash of their public keys. Instead of s-expressions, AOIS uses XML signed documents, in the form of SAML assertions [21], to convey identity, role and property assignments. As in SPKI, delegation is possible if the delegating principal issues a certificate whose subject is a name defined by another, trusted, principal. The latter can successively issue other certificates to assign other principals (public keys) to its local name. In this sense, local names act as distributed roles [14].

Finally, the extraction of a digest for each search result is required to avoid the presentation of duplicate results to the user. This feature is provided by a Java implementation of a hash function [24].

5 Testing

Practical tests on the first prototype of the AOIS system were done installing the system in different labs and offices of our department asking some students and colleagues to use it for sharing and exchanging information. Moreover, we tested the system setting some computers of a Lab with different access policies and distributing information on their repositories providing, in some cases, different copies of the same information on different computers. The tests covered with success all the system features and the searching and pushing of information satisfied our expectations.

Moreover, a part of the experimentation was oriented to compare the results of the searching and pushing operations based on the use of topic ontologies with the ones based only on the use of keywords and what happened is that: i) the use of topic ontologies increases the number of results, but very few were of no interest for the users if, in particular, the users chose a good set of top terms.

Up to now, we do not perform a numeric analysis of the results, but only a qualitative analysis derived from a discussion with the people that used the system. The main result is that usually the quality of search and pushing operations mainly depends on an appropriate set of top keywords. Therefore, the goodness of an ontology usually does not depends on the keywords extracted from the information of the repository, but mainly depends on an appropriate initial set of top keywords and then by the introduction of the other appropriate keywords coming from the queries of remote users.

6 Conclusions

In this paper, we presented a peer-to-peer multi-agent system, called AOIS (Agents and Ontology based Information Sharing), that supports the sharing of information among a community of users connected through the Internet. AOIS is an evolution of a previous system [15], called RAIS (Remote Assistant for Information Sharing), that performed a similar task, but was implemented by using a different search technology (i.e., Google Desktop Search) and did not take advantage of topic ontologies for the search of information. The first prototypes of the AOIS system used the JXTA protocols to provide the peer-to-peer features useful to support the interaction among remote users. However, JXTA is not used in the most known and used peer-to-peer applications and so it is suitable to test the features of prototypes, but it is unsuitable for developing real application. Therefore, the last AOIS implementation is based on BitTorrent, one of the most widespread and solid file-sharing platform, which can be configured and extended to work in a completely decentralized fashion.

AOIS derives a large part of its features from the systems for information sharing described in the related work section. However, it offers a new feature that seems to improve the quality of search and pushing operations: the creation of a topic ontology through the use of a set of initial terms (i.e., the top terms), its automatic extension through the information maintained by the user, the possibility of controlling the semantic distance from the top terms and the terms automatically added, and, in particular, the possibility of using the terms contained in the queries of the other users for refining the ontology, allow the construction of high quality ontologies. Then, a topic ontology can be customized by each user, but taking into account of the implicit suggestions of the other users of the community, Moreover, its implementation based on some well-known software tools guarantees good performance and reliability.

The first prototypes of the AOIS system was experimented in some "artificial" communities involving researchers and students of our University and the results of the experimentation encouraged us in the further development. The introduction of BitTorrent in its last prototype will allow us to plan a real and large experimentation thinking to the hundreds of millions of users that already share documents though the BitTorrent platform and may be interested in using such a system.

The current implementation of the system maintains in the remote user profiles all the queries she/he did. Often the information retrieved through some old queries might be not yet of interest for the remote user. Therefore, we are working on a more sophisticated technique for managing remote user profiles: all the queries are stored together with the time they were executed; every day the IP checks the remote user profiles and for all the queries that are older than a fixed duration (e.g., a week), it asks the PA about the interest of its user in maintaining such queries and refreshes the execution time for all the queries for which it receives a positive acknowledge.

The creation of topic ontologies may be a difficult activity because it requires the identification of an appropriate set of top terms and its completion through the use of an appropriate set of information. Therefore, the possibility of using the topic ontologies built by other users may be an important feature of the system. In fact, we are working to the possibility that PA agents can require some topic ontologies to other PA agents and then either directly use them for driving the search or build new topic ontologies by merging them with the local topic ontologies.

Beyond the definition of the top terms and of the maximum semantic distance between terms, users have not the possibility of managing the topic ontologies. But this would be a very important feature in the future, when the system will allow the use of topic ontologies defined by other users and the merging among different topic ontologies. In the current version of the system the topic ontologies are also saved as OWL files because the search ontology Nutch plugin requires an OWL file for proving ontology based search. Therefore, users may manipulate topic ontologies by using one of the available tools for manipulating OWL ontologies (e.g., Protégé [24]). However, in the OWL view of the topic ontologies there is not information about the top terms. Therefore, we are developing a very simple graphical tools (based on the use of the Jung software library [12]) that: i) shows the graph defining an ontology, ii) distinguishes top terms from the other terms, iii) distinguishes the different kinds of semantic relationships among terms, iv) allows the introduction of new terms and the deletion of existing terms, v) allows the introduction and the deletion of the "top" attribute to any term, and vi) allow the modification of the maximum semantic distance (when such a distance is reduced, the tool removes all the terms that do not satisfy the new constraint).

Acknowledgments. This work is partially supported by the Italian Ministry MIUR (Ministero dell'Istruzione, dell'Università e della Ricerca).

References

1. Apache Foundation, Nutch software (2011), http://nutch.apache.org
2. Beagle Team, Beagle software (2011), http://beagle-project.org
3. Bellifemine, F., Poggi, A., Rimassa, G.: Developing multi agent systems with a FIPA-compliant agent framework. Software Practice & Experience 31, 103–128 (2001)
4. Bellifemine, F., Caire, G., Poggi, A., Rimassa, G.: JADE: a Software Framework for Developing Multi-Agent Applications. Lessons Learned. Information and Software Technology Journal 50, 10–21 (2008)
5. Chen, J.R., Wolf, S.R., Wragg, S.D.: A Distributed Multi-Agent System for Collaborative Information Management and Sharing. In: Proc. of the 9th ACM International Conference on Information and Knowledge Management, pp. 382–388 (2000)
6. Cohen, B.: Incentives build robustness in BitTorrent. In: Proceedings of the First Workshop on Economics of Peer-to-Peer Systems, Berkeley, CA (2003)

7. Ellison, C., Frantz, B., Lampson, B., Rivest, R., Thomas, B., Ylonen, T.: SPKI Certificate Theory. RFC 2693 (1999)
8. FIPA Consortium, FIPA Specifications (2011), http://www.fipa.org
9. Gong, L.: JXTA: A network programming environment. IEEE Internet Computing 5(3), 88–95 (2001)
10. Google, About Google Desktop Search software (2011), http://desktop.google.com
11. Hotho, A., Jäschke, R., Schmitz, C., Stumme, G.: Information Retrieval in Folksonomies: Search and Ranking. In: Sure, Y., Domingue, J. (eds.) ESWC 2006. LNCS, vol. 4011, pp. 411–426. Springer, Heidelberg (2006)
12. Jung Team, Jung software (2011), http://jung.sourceforge.net
13. Klusch, M.: Information Agent Technology for the Internet: A survey. Data & Knowledge Engineering 36(3), 337–372 (2001)
14. Li, N., Mitchell, J.M.: RT. A Role-based Trust-management Framework. In: Proc. of the Third DARPA Information Survivability Conference and Exposition (DISCEX III), Washington, DC, pp. 201–212 (2003)
15. Mari, M., Poggi, A., Tomaiuolo, M., Turci, P.: Enhancing Information Sharing Through Agents. In: Kolp, M., Henderson-Sellers, B., Mouratidis, H., Garcia, A., Ghose, A.K., Bresciani, P. (eds.) AOIS 2006. LNCS (LNAI), vol. 4898, pp. 202–211. Springer, Heidelberg (2008)
16. Maymounkov, P., Mazières, D.: Kademlia: A Peer-to-Peer Information System Based on the XOR Metric. In: Druschel, P., Kaashoek, M.F., Rowstron, A. (eds.) IPTPS 2002. LNCS, vol. 2429, pp. 53–65. Springer, Heidelberg (2002)
17. Miller, G.A.: WordNet: A Lexical Database for English. Communications of the ACM 38(11), 39–41 (1995)
18. Mine, T., Matsuno, D., Kogo, A., Amamiya, M.: Design and Implementation of Agent Community Based Peer-to-Peer Information Retrieval Method. In: Klusch, M., Ossowski, S., Kashyap, V., Unland, R. (eds.) CIA 2004. LNCS (LNAI), vol. 3191, pp. 31–46. Springer, Heidelberg (2004)
19. Montassier, G., Cholez, T., Doyen, G., Khatoun, R., Chrisment, I., Festor, O.: Content pollution quantification in large P2P networks: A measurement study on KAD. In: Proc. of the IEEE International Conference on Peer-to-Peer Computing, Kyoto, Japan, pp. 30–33 (2011)
20. Riaño, D., Moreno, A., Isern, D., Bocio, J., Sánchez, D., Jiménez, L.: Knowledge Exploitation from the Web. In: Karagiannis, D., Reimer, U. (eds.) PAKM 2004. LNCS (LNAI), vol. 3336, pp. 175–185. Springer, Heidelberg (2004)
21. OASIS, SAML specifications (2011), http://saml.xml.org
22. Parameswaran, M., Susarla, A., Whinston, A.B.: P2P Networking: An Information-Sharing Alternative. Computer 34(7), 31–38 (2001)
23. Poggi, A., Tomaiuolo, M.: A Multi-Agent System for Information Semantic Sharing. In: Proc. of the 5th International Workshop on New Challenges in Distributed Information Filtering and Retrieval, Palermo, Italy (2011)
24. Princeton Universty, Wordnet (2011), http://wordnet.princeton.edu
25. Rivest, R.L.: The MD5 Message Digest Algorithm. Internet RFC 1321 (1992)
26. Salter, J., Antonopoulos, N.: CinemaScreen Recommender Agent: Combining Collaborative and Content-Based Filtering. IEEE Intelligent Systems 21(1), 35–41 (2006)

27. Sánchez, D., Isern, D., Moreno, A.: Integrated Agent-Based Approach for Ontology-Driven Web Filtering. In: Gabrys, B., Howlett, R.J., Jain, L.C. (eds.) KES 2006. LNCS (LNAI), vol. 4253, pp. 758–765. Springer, Heidelberg (2006)
28. Southern Methodist University, JAWS software (2011), http://lyle.smu.edu/~tspell/jaws
29. Stanford University, Protégé software (2011), http://protege.stanford.edu
30. Tang, C., Xu, Z., Dwarkadas, S.: Peer-to-peer information retrieval using self-organizing semantic overlay networks. In: Proc. of ACM SIGCOMM, pp. 175–186 (2003)
31. Telecom Italia, JADE software (2012), http://jade.tilab.com
32. Vuze BitTorrent Client (2012), http://vuze.sourceforge.net
33. W3C Consortium, OWL 2 Web Ontology Language Overview (2009), http://www.w3.org/TR/owl2-overview
34. Wiemer-Hastings, P.M.: How Latent is Latent Semantic Analysis? In: Proc. of the Sixteenth International Joint Conference on Artificial Intelligence (IJCAI 1999), pp. 932–941 (1999)

A Decisional Multi-Agent Framework for Automatic Supply Chain Arrangement

Luca Greco, Liliana Lo Presti, Agnese Augello, Giuseppe Lo Re, Marco La Cascia, and Salvatore Gaglio

Abstract. In this work, a multi-agent system (MAS) for supply chain dynamic configuration is proposed. The brain of each agent is composed of a Bayesian Decision Network (BDN); this choice allows the agent for taking the best decisions estimating benefits and potential risks of different strategies, analyzing and managing uncertain information about the collaborating companies. Each agent collects information about customer's orders and current market prices, and analyzes previous experiences of collaborations with trading partners. The agent therefore performs a probabilistic inferential reasoning to filter information modeled in its knowledge base in order to achieve the best performance in the supply chain organization.

1 Introduction

In this paper, we present a decision support system for companies involved in the organization of a supply chain. In a supply chain, a group of suppliers and customers collaborate to provide services/products with a quality greater than the one a single partner could provide alone. Within this network, persons and hardware/software systems work together sharing information and services in order to provide a product that can satisfy the final customer needs.In such a complex environment decisions must be made quickly, analyzing and sharing several information with multiple actors [1].

Luca Greco · Liliana Lo Presti · Giuseppe Lo Re ·
Marco La Cascia · Salvatore Gaglio
DICGIM, University of Palermo,
Viale delle Scienze, Edificio 6 90128, Palermo, Italy
e-mail: {greco,lopresti}@dicgim.unipa.it,
{giuseppe.lore,marco.lacascia,salvatore.gaglio}@unipa.it

Agnese Augello
ICAR Italian National Research Council, Viale delle Scienze
e-mail: augello@pa.icar.cnr.it

C. Lai et al. (Eds.): New Challenges in Distributed Inf. Filtering and Retrieval, SCI 439, pp. 215–232.
springerlink.com

A supply chain management system includes several entities: the different companies involved in the supply chain and, for each company, different entities specialized in the accomplishment of specific business tasks. Such scenario makes urgent the realization of new tools for effectively retrieving, filtering, sharing and using the information flowing in the network of suppliers/customers. Such tools should be used to continuously monitor business processes and collaborations established within the group; moreover, they have to be conceived so to limit the dependencies among companies guaranteeing each company could work independently of each other. Supply chains are constantly subject to unpredictable market dynamics, and in particular to continuous changes in prices and in commercial partnerships, which may become more or less reliable. The uncertainty that characterizes these changes can affect the supply chain performance and should be properly handled [2][3].

Multi-agent systems may be particularly useful for modeling supply chain dynamics [4]. The entities involved within a supply chain can be represented by agents able to perform actions and to make autonomous decisions in order to meet their goals [5][6]. Supply chain organization can be therefore implemented as a distributed process where multiple agents apply their own retrieval and filtering capabilities.

In this work, we envision a system where the supply chain can be automatically organized maximizing the utility of the whole group of collaborating partners. The success of the established collaborations depends on the capability of each company and of the whole group to adapt to the changes in the environment they work within. Adapting their strategies and behaviors, the whole group of partners is able to exploit new solutions and configurations that can assure the production of high quality products and can guarantee a profit for each partner. The success of the collaboration enables the success of each company and, therefore, the companies are motivated to adopt strategies to increase their profit while easing the establishment of new collaborations. The knowledge, relations and processes within the group of companies results from continuous negotiations among the partners in order to reach a consensus about their availability to join the supply chain. Once a consensus has been reached, a supply chain has been organized. Then, several suppliers and customers will collaborate to maximize their profit and satisfy the customer needs while providing a high quality product at a convenient price. Given a set of companies, the problem of organizing a supply chain has exponential complexity, being necessary to evaluate several kind of potential supply chains for choosing the one that satisfy all the constraints previously described. Solving such problem requires both techniques for retrieving the information and for filtering and using it in the most convenient way. Therefore, we propose a method for the supply chain creation with the goal of maximizing the utility of each single company and realize the best collaboration within a group of companies that agree to work together.

The system we propose is made of a community of intelligent agents able to provide support in supply chain decision processes. Each agent is responsible for decision-making processes relating to a particular company. To accomplish this goal each agent has to retrieve information necessary for decision making, incorporating them in its own knowledge base.

Decision making is not a simple activity but a process leading to the analysis of several variables, often characterized by uncertainty, and the selection of different actions among several alternatives. For this reason, the brain of each agent has been modeled by means of a Bayesian Decision Network (BDN)[7]; this choice allows the agent for analyzing uncertain information and estimating the benefits and the potential risks of different decisional strategies. A preliminary version of this work has been presented in [8].

The paper is organized as follows. In Section 2 some related works are discussed and in Section 3 the proposed system is described, while in Section 4 a case study is reported; finally Section 5 reports conclusions about the proposed system and future works.

2 Related Works

Multi agent systems (MASs) provide an appropriate infrastructure for supporting collaborations among geographically distributed supply chain decision-makers [9] [10]. Moreover, the agent paradigm is a natural metaphor for supply chain organizations, because it allows for easily modeling supply chain member features (decisional autonomy, social ability to establish agreements with the other companies, reactivity to the market, but also proactiveness)[11].

There exists a large literature on this subject, an interesting review is reported by [4]. Agent systems can be used for supply chain management or design purposes [11]. Here we analyze issues related to the dynamic configuration of supply chains.

In this context, a team of researchers from the e-Supply Chain Management Lab at Carnegie Mellon University and the Swedish Institute of Computer Science (SICS) has defined the rules of a competition called TAC SCM (Trading Agent Competition - Supply Chain Management Game), aimed to capture many of the challenges involved in supporting dynamic supply chain practices [12].

Authors of [13] propose a multi agent system which allows for a dynamic reconfiguration of the supply chain, managing scenarios where suppliers, prices and customers demands may change over time.

In [14], a machine learning algorithm based on decision tree building allows for the choice of the best node at each stage of the supply network analyzing the combination of parameters such as price, lead-time, quantity, etc. In particular, at each node, an agent collects information about the upstream nodes, filtering the information to extract necessary training examples for a learning module. The learning module, therefore, extracts from the training examples a set of decision rules which are used by a dispatcher to identify the best choice of a node.

In previous works [17] [18] [19], a set of agents organizes a coalition to execute a complex task based on the contract net protocol [16]. In this kind of system, a customer can specify the good he wants to buy as well as the maximum price he is willing to pay for it. The agent system finds the user willing to fulfill the customer's order at the lowest price. During the negotiation, the agents establish contracts that specify and regulate the agents' interactions allowing the required task to be distributed among a group of agents. In this sense, the task must be precise and hierarchical in nature, i.e. the task can be broken down into mutually independent subtasks. Our framework is similar in spirit to the previous works, but enables each agent to reasoning in a probabilistic way by means of a BDN modeling the supply-chain organization process. A supplier selection process involves subjective, imprecise and uncertain information that must be translated into quantitative data for decision making [21]. In [15] and [21] fuzzy sets theory is used to translate this vague information into quantitative data in order to define supplier selection criteria. [2],[3], [22] focus on uncertainty issues in the organization of supply chains. The authors of [2] study the effect of uncertain customer demand, supplier capacity, and supplier's capacity utilization; they employ an agent-based simulation to evaluate two different adaptive coordination strategies. In [3] different coordination and information sharing techniques are analyzed in order to understand which combination is the most effective in managing uncertainty. In particular, in [22], a theoretical model, based on an extension of Bayesian Networks models, is used to formalize supply chain agents' interactions during an order fulfillment process. The direct supply-demand relationships between pairs of agents are modeled as directed causal links, because the failure of a supplier to fulfill its commitments may affect the commitment progress of his customers. The information sharing between agents is modeled as belief propagation. The extended Bayesian Belief Network model proposed by the authors allows the agents to perform strategic actions, such as dynamically select or switch the suppliers, or take decisions to cancel a commitment based on its related expected utility function.

As in [22], we model the main variables of the domain and their causal relations, but exploiting the advantages of Bayesian Decision Network (BDN) models. Each agent of the chain has its own BDN to represent its beliefs about the reliability of commercial partners, based on trade relationships established during the past business experience. The information arising from each network is used for configuring the entire supply chain.

3 Proposed System

In this paper, we make the assumption that each company is represented by an agent, and we focus on the information retrieval and filtering process performed by each agent to organize a supply chain. More complex architecture can be used to model the business process within each company, but we assume there is an agent that represents the entire company and is involved in the supply chain organization process. Considering the characteristics of the system we are envisioning, we believe a multi-agent system is the proper tool for its implementation. Indeed, the agent design paradigm permits to realize several autonomous components of a complex system; moreover, agents can make easier the exchange of information and services to realize virtual collaboration across the network of companies. Agents are able to take actions autonomously and independently one each other by means of their social capabilities and pro-activeness; this contributes to model the distributed and dynamic environment supply chains must be realized in.

Figure 1 shows how the supply chain creation process is triggered. We assume each agent in the supply chain analyses the information provided by the informative system of the company it works on behalf of. Such informative system constantly updates the agent about the current available resources, the productive capabilities, the time required for developing business processes and other useful information. The supply chain decisional process starts when a new order arrives to a company. The agent responsible for that company will then trigger a decision-making process aimed at the supply chain building. This process leads in turn to the creation of sub chains, performed by other agents involved in the fulfillment of the order. To create its own sub-chain each agent queries the informative system of its company. The brain of each agent is composed of a Bayesian Decision Network (BDN); this choice allows the agent to take the best decision estimating the benefits and the potential risks of different strategies, analyzing and managing uncertain information about enterprise data. Moreover the agent's knowledge base can be constantly and dynamically learned and updated by means of observations on data managed by the informative systems of the companies.

An agent representing a certain company in the supply chain can act as supplier if the company is a provider of some material, or as supplier/customer if the business activity of the company requires other products or raw materials from other companies in the chain. In the first case, the agent has a passive role in the supply chain creation process and its decisional strategies will regard price negotiations to join the chain. In the second case, the agent has an active role, because it will create a sub-chain to manufacture its own product. We also assume that, whenever the agent needs to buy a certain material from the other agents, it will buy the necessary quantity from a single supplier. This hypothesis does not limit the applicability of our system because the system could be easily extended considering as possible suppliers combinations of agents that jointly fulfill the order based on their capabilities and available resources.

Moreover, to account for the changes of the market and environment conditions and/or to changes of the business strategies adopted by each company within the group, it is required to exploit new kind of collaborations among the companies over time. Each company automatically decides if and when to assume the role of customer/supplier for the other partners across the supply chain.

3.1 Dynamic Supply Chain Organization

During the supply chain organization process, each agent can receive an invitation to join the chain as supplier and, based on its actual resources, productive capabilities and economical convenience, it can decide if joining the chain or no. To collaborate in the chain, the agent can ask other agents for products/services it needs for its own business process. In this case, it acts as customer and invites other agents to join the chain for satisfying a certain order. In practice, before joining the supply chain, the agent needs to organize a sub-chain for its own business process. Once an agent knows it can join the supply chain, it replies to the invitation informing about its

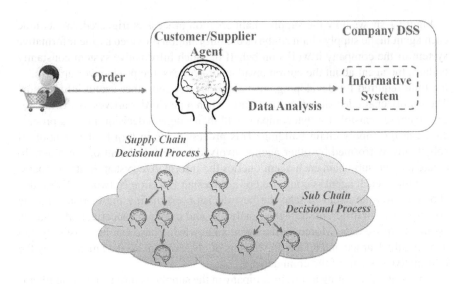

Fig. 1 In the proposed system, a set of agents collaborates to fulfill the customer's order. The agent receiving the order triggers the supply chain organization process that is developed hierarchically within the group of collaborating companies.

availability and the conditions it wants to impose for being part of the collaboration (for example price and temporal conditions). Then, it waits to be notified if it has been selected as partner of the supply chain and, in this case, it starts the collaboration. Therefore, a supply chain is the result of a set of negotiations among the agents belonging to the same group. All the agents communicate costs, quantities, times and modalities needed for the supply chain organization. To reach a consensus, two different kinds of information flows are used across the agent network. As depicted in figure 2, the top-down information flow represents the invitations sent to agents for being part of the supply chain, while the bottom-up information flow represents the information flowing from suppliers to customer about their conditions to join the supply chain.

During the negotiations, all the agents are in competition one each other; their behavior can be oriented to maximizing their business volume constrained by the quality of the final product, their productivity capability and the minimum profit they want to get. The supply chain can be built choosing all the agents that, with the highest probability, could assure the success of the final supply chain and would collaborate to satisfy the final customer's order. The problem of automatically organize the supply chain has an exponential complexity being necessary to evaluate all the possible supply chain hypotheses. In the following, we assume that the entire supply chain can be modeled as a tree; at each level, a sub-tree represents a sub-chain. In our formulation, at each node of the tree an agent provides a particular goods or service needed at higher levels to provide the product required by the customer. However, the production of this goods/services can require the cooperation of other agents. Therefore it can be necessary to establish a set of collaborations with other agents, i.e. a new sub-chain. The organization of the entire supply chain reduces to recursively organize each sub-chain as showed in figure 2. The problem of organizing a supply chain is therefore addressed by dividing it into sub-problems of lower complexity. The entire supply chain can be organized considering sub-optimal solutions at each node of the tree. Whilst in general the solution will not be globally optimal, under the assumption of independence of the sub-tree, the final solution will be optimal. This assumption does not limit the applicability of our system because it is reasonable to assume that at each node of the supply chain independent business processes would be developed.

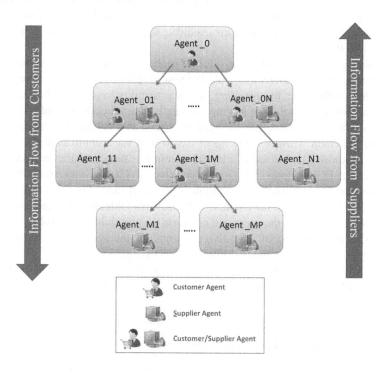

Fig. 2 A Supply Chain can be organized hierarchically. Each agent can build a sub-chain to develop its own business process. Each agent behaves as supplier towards the higher level in the tree, and as customer for the agents at the lower level in the tree. The image shows the two different information flows within the agent network.

3.2 Agent-Based Chain Organization

Algorithm 1 reports the pseudo-code for the finite state machine adopted to implement the agents in our system. In capital letters we specify behaviors triggered and used by the agents to pursue their goals. In cursive capital letters we specify messages sent/received by the agent.

First, the agent has to collect or update its beliefs considering the success/failure of the supply-chains it joined.

Once it receives an invitation to participate to a supply-chain, then the agent checks if, based on its adopted marketing strategies and its current capabilities, it is convenient or not to join the chain (CHECK_CONVENIENCE). In our implementation, we assumed the agents always find convenient to join the chain.

When BUSINESS_PROCESS_ORGANIZATION is invoked, the agent analyzes the current stocks and plans the resource allocation necessary to satisfy the customer's request. It also selects the list of suppliers to contact for organizing its own sub-chain. In the last case, the agent is going to expand the whole supply-chain tree by adding a sub-tree corresponding to its own sub-chain. To organize such sub-chain, the agent sends invitations to the candidate suppliers and waits for their availability to join the chain.

Once the agent receives at least an offer for each of the goods it needs to buy, then the agent may adopt the BDN to choose the best supplier combinations to organize the sub-chain. Otherwise, the agent has not the possibility to organize the chain and send to the customer a denial of joining the supply chain. In case the sub-chain has been successfully organized, then the agent may send an acceptance to the customer in joining the chain. However, the agent has to wait for a confirmation from the customer to know if it has been selected to join the supply chain. In particular, similarly to the FGP (Finite-Time Guarantee Protocol) [20], in our framework each agent waits for the confirmation for a limited period of time. After this time, if the agent is not selected to join the chain, it will make available the allocated resources for other bids. In this way, the agent will not loose the possibility to join other chains.

Only after the receiving of a positive confirmation, the agent may acknowledge all its suppliers and starting the real PRODUCTION_PROCESS.

In our implementation, tested on simulations, we did not implemented the PRODUCTION_PROCESS, but we updated the virtual capabilities of the agent to join future supply chains.

3.3 Supply Chain Decisional Process

The supply chain organization requires the selection of agents whose cooperation can ensure the success of the entire supply chain. It is necessary to adopt strategies that take into account the uncertainty of the environment in which the agents work. In fact, the establishment of a supply chain is not a deterministic process. Factors such as delays in delivery – for example due to an excessive geographical distance of the companies the agents represent –, the reliability of a supplier and consequently the failure to meet his commitments could determine a failure for the supply chain.

Our system takes into account such factors by assigning a degree of uncertainty to the choice of agents in the supply chain: each agent can join the supply chain with a certain probability.

Algorithm 1. Finite State Machine for Agent

1: UPDATE/COLLECT BELIEFS
2: Wait for any *INVITATION* to join/organize the supply-chain;
3: **if** *INVITATION* has been received **then**
4:     CHECK_CONVENIENCE of joining the chain
5:     **if** positive CONVENIENCE **then**
6:         BUSINESS_PROCESS_ORGANIZATION:
           Analyze stock, select quantities and candidate suppliers
7:         **EXPAND_TREE:**
8:         SEND *INVITATIONS* to candidate suppliers
9:         Wait for all the *RESPONSES* (within a temporal window)
10:        CHECK_RESPONSES:
11:        **if** at least a positive response for each necessary material has been received **then**
12:            ORGANIZE_SUB-CHAIN:
               Use BDN to compute the best suppliers organization
13:            **if** SUB-CHAIN has been organized **then**
14:                DECIDE_PRICE
15:                SEND *ACCEPTANCE* and *PRICE* to Customer
16:                Wait for *CONFIRMATION* of the Customer
17:                **if** positive *CONFIRMATION*
18:                    SEND *POSITIVE_CONFIRMATIONS* to selected suppliers
19:                    SEND *NEGATIVE_CONFIRMATIONS* to discarded suppliers
20:                    START_PRODUCTION_PROCESS
21:                **else**
22:                    SEND *NEGATIVE_CONFIRMATIONS* to suppliers
23:                **endif**
24:            **else**
25:                SEND *DENIAL*
26:            **endif**
27:        **else**
28:            SEND *DENIAL*
29:        **endif**
30:    **else**
31:        SEND *DENIAL*
32:    **end if**
33: **end if**
34: goto line 1

To take into account the uncertainty of the business process, in our system each
agent adopts a Bayesian Decision Network (BDN) to represent explicitly considerations about cost-benefits associated with each strategy in the decision-making
process. Our system takes into account the uncertainty of the process, representing
these factors in an appropriate model. In particular the knowledge base of each agent
is modeled by a Bayesian Decision Network (BDN), an extension of a Bayesian
Belief Network which permits to represent explicitly considerations about cost-
benefits associated with each strategy in a decision-making process. BDNs allow for

considering the agent state, its possible actions and the associated utility. For all these properties, BDNs can support humans in taking decisions, simulating, analyzing or predicting scenarios. Their use is particularly appropriated when timely decisions must be taken in complex and dynamic domains, and where multiple actors are present. Moreover, proper learning algorithms allow for a dynamic generation and updating of agents knowledge models. These models are used in this context to manage strategic commercial decision policies and arrange the supply chain.

Through the combination of probabilistic reasoning with the utilities associated with the decisions, the BDNs are a valid tool to choose the decision that maximizes the expected utility.

In our system, each agent can assign a degree of uncertainty to the success or failure of a particular configuration for a supply chain. To develop its business process, an agent retrieves the information about all the suppliers available to join its own chain and reasons on the collected information; then, by means of a BDN, the agent filters the suppliers and retains only those able to organize the best sub-chain.

The probabilities of the network may be a priori known or on-line learnt based on strategies chosen by each agent. In this sense, several strategies can be adopted, especially depending on the type of market. Business decisions can be taken in relation to parameters of convenience, as generally done in case of wide consumer products, or considering other factors such as the prestige, the competitiveness or the brand of the potential suppliers, as happens in a market of luxury or highly differentiated products. In this case, in establishing a trade agreement, the agent will consider factors such as price, delivery terms and payment, product quality and business partners reliability. As an example, cost leadership strategies focus on acquiring highest quality raw materials at the lowest price. Instead, in a market of luxury or highly differentiated products, decisions may be taken considering other factors such as the prestige, the competitiveness or the brand of the potential suppliers. Different strategies can be appropriately modeled in the BDN.

3.3.1 Supply Chain Decision Network

Figure 3 shows an example of decision network used by each customer agent to organize its own sub-chain.

The strategy of each agent is to compare the reliability and the price proposal of different suppliers. In particular, to take into account the reliability of an agent to meet its commitments when involved in a collaboration, each agent is associated with a "reputation". Based on its past experience or on specific adopted strategies, each agent associates each supplier with a certain reputation that it can adapt over time. In practice, the reputation represents how much the agent trust in its suppliers. For example, it is possible to on-line learn such value by measuring the number of times a supplier has fulfilled its commitments and/or evaluating the quality of the collaborations in which the supplier has been involved (by analyzing a set of parameters such as the product/service quality and time delivery), or considering the supplier degree of specialization or expertise in the field.

In the specific example, it is assumed that an agent needs to buy two different types of materials to manufacture its product. Each material is associated with a list of possible supplier agents. The reputation node (which shows the reputation associated with the supplier – *low, medium, high, not available*) and the offer node are conditioned on the choice of the supplier; the offer node is modeled by a deterministic node (*SupplierMaterialOffer*) representing the cost of the offer proposed by the supplier (*sufficient, fair, good, not available*). This value may be determined by comparing the received offer to the market price and/or evaluating the quality and characteristics of the offered products.

The reputation associated with a supplier agent influences the commercial transaction to acquire the specific material, represented by the node *TransactionStatus*. The reputation and the offer jointly determine the utility associated with the transaction between the customer agent and the selected supplier agent (utility node *TransactionUtility*). The value of the expected utility is the expected value associated with the offers according to the reputation probability distribution. The Transaction utility nodes for all the needed materials are used together to determine the utility of the entire sub-chain. The Transaction probabilistic nodes for raw materials, instead, influence the probabilistic node *Supplychain*, which expresses the probability that a supply chain can be successfully established.

Through the proposed BDN, the decision analysis is performed directly comparing the utility values corresponding to different choices ensuring the success of the supply chain. We stress the BDN parameters implement the strategy that each agent

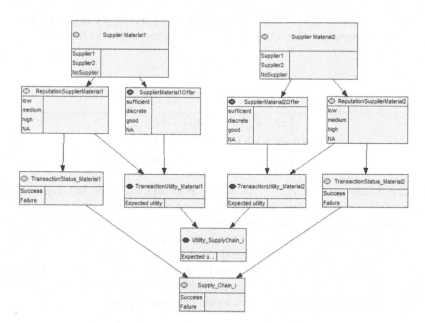

Fig. 3 Agent Decision Network for selecting the suppliers that maximizes the utility associated with the supply chain. In the figure, only two kind of materials have to be bought.

wants to adopt when assembling its supply chain. Parameters of the utility nodes for each agent can be devised by a knowledge engineer according to the sale manager guidelines in order to represent the strategy for the company the agent works on behalf of.

3.4 Dynamic Behavior of Agents over Time

Although the agents may adopt different strategies, in our formulation we assume each agent chooses whether to participate in the organization of a supply chain based on the available resources. If it decides to join the chain, the agent sets the selling price taking into account constraints related to the minimum profit it wants to get from the transaction, and its experience in previous negotiations. Therefore, in our system each agent decides whether to increase or decrease the selling price based on the outcome of the offers in previous transactions within a given time window. If the agent has not been selected for joining a supply chain at a certain selling price then, at the next negotiation, it decreases the selling price (constrained by the costs necessary for production). Conversely, if its previous offers have been accepted at a certain price, it tries to slightly increase the selling price in order to maximize its profit. As a consequence, in this scenario, the agents adopt a lower price strategy, where prices tend to decrease and agents become as more competitive as possible.

Let P_m be the minimum selling price to cover production costs and ensure a minimum profit, and P_M the maximum selling price the agent knows it is risky to sell to (this price could be set by the customer of the agent). The selling price varies according to the number of times k the agent has joined a supply chain in the time window T. In particular, the selling price P varies according to the following:

$$P = \max(P_m, P^*) \qquad (1)$$

where

$$P^* = \begin{cases} \min(P_{t-1} + \Delta P; P_M) & \text{if } k \geq \tau \\ P_{t-1} - \Delta P & \text{if } k \leq \gamma \\ P_{t-1} & \text{otherwise.} \end{cases} \qquad (2)$$

In the previous formula, P_{t-1} is the selling price offered at the last negotiation, while ΔP is a priori known value representing how much the selling price is increased/decreased. τ represents the threshold value to determine after how many successful transactions in T the price would be increased; conversely, the threshold γ is the maximum value of k for which the price should be reduced.

Although more complex strategies could be applied, the one just described allows agents for dynamically adapting to the environment in which they operate favoring a free competition among the partners within the same group.

4   Experimental Results

To evaluate the system, we considered the simple case where the supply chain can be represented by a binary tree and each agent can have no more than two different suppliers for each material it needs to buy. As discussed in Section 3.3.1, it is possible to learn agent reputation considering the success of the negotiations among agents through the time. Here, for the sake of demonstrating our approach, i.e. adopting a BDN for taking decisions about the supply chain organization, we assume agent reputations are constant across time and study how the negotiations are carried on among the agents. We empirically demonstrate that, as consequence of the supply chain organization outcome and, therefore, of the decisions taken by each agent, at each negotiation the agents change the offered price in order to join more supply chains as possible.

To implement our Multiple-Agent System, we adopted the Java Agent DEvelopment Framework (JADE) [23], that simplifies the development of distributed agent-based systems. The BDN has been implemented by GeNIe [24], that offers a simple environment to design decision-theoretic methods. We also implemented simple wrapper classes the agents can use to interface with GeNIe.

We simulated a network composed of 21 different agents as showed in figure 4. We are assuming the agents have to buy only two materials, and for each one they would contact only two suppliers. Of course, this configuration has meant for demonstrating the validity of the proposed approach, but in real cases more complex scenarios can be handled. We also assume each agent offers its product at a certain price computed based on a certain strategy, i.e. the one presented in Sec. 3.4.

Let $\{S_i\}_{i=1}^5$ be the set of agents that can act both as supplier and customer within the supply-chain; therefore, they need to organize their own sub-chains. Let $\{A_j\}_{j=1}^{16}$ be the set of agents that have the role of suppliers within the supply chain; these agents do not need to organize any supply chain for their business processes. Let $\{P_k\}_{k=1}^{10}$ represent the products that are sold/bought within the agent network to assemble the supply chain.

In figure 4, the agent S_1 has the role of supplier to the final real customer and assembles the main supply chain for satisfying the incoming orders. Agent S_1 provides the product P and needs to buy two different kinds of materials: P_1 and P_2. Then, it sends an invitation to join the chain to its own suppliers, that are S_2 and S_3 for the material P_1 and S_4 and S_5 for the material P_2. To provide the product P_2, the agent S_2 needs the materials P_3 and P_4 and, therefore, it contacts the corresponding suppliers: A_1 and A_2 for P_3, and A_3 and A_4 for P_4. This kind of hierarchical structure is repeated for each agent at the second level. In this specific scenario, only the agents at the first and the second level of the tree decide to organize a supply chain and, therefore, they use the proposed decision network for choosing the configuration of suppliers that would guarantee their maximum utility.

4.1   An Example of Supply Chain Decision Making

The supply chain organization requires a set of negotiations among the agents to exploit possible collaborations and choosing the one that maximizes the agent utility. A negotiation can be modeled as an exchange of messages among agents that communicate their availability to join the chain at certain conditions, i.e. prices, quantities and time.

The selection of the agents joining the chain is done considering the utility associated with the supply chain; based on the value of reputation and cost, it is possible that a supplier offering an higher price is selected because of its reputation. For example, let consider the scenario represented in table 1. Two suppliers with the same probability distribution of the reputation (*Reputation low, medium, high and NA*) are selling the same material at two different prices that the agent considers discrete and good respectively. In this case, the BDN permits the agent to select the most convenient price. This can be easily seen comparing the utilities associated to each transaction.

In the case represented in table 2, two suppliers with different probability distribution of the reputation are selling the same material at two different prices that the agent classifies as discrete and sufficient respectively. In this case, whilst the first supplier is offering the most convenient price, the agent chooses the supplier with the better reputation distribution.

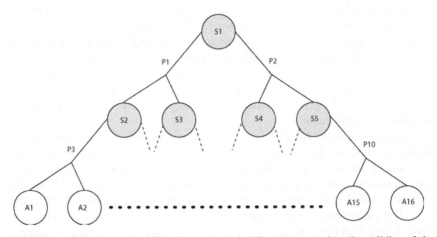

Fig. 4 The figure shows the environment simulated for demonstrating the validity of the proposed system. Each node in the tree represents an agent. Only five agents need to build sub-chains. To organize the supply chain, agent S_1 needs to buy products P_1 and P_2. Therefore, it has to choose between S_2 and S_3 and between S_4 and S_5. On the other hand, S_2 needs to organize its own sub-chain before raising an offer to S_1. It needs to buy P_3 from the agent A_1 or A_2 and the material P_4 from the agent A_3 or A_4. Every agent of second level works similarly.

Table 1 Scenario 1 – Supplier selection in case of same reputation distribution and different offers.

Nodes	Reputation Low	Reputation Medium	Reputation High	Reputation NA	Offer	TransactionUtility
Supplier 1	0.33	0.33	0.33	0	discrete	6
Supplier 2	0.33	0.33	0.33	0	good	7

Table 2 Scenario 2 – Supplier selection in case of different reputation distributions and offers.

Nodes	Reputation Low	Reputation Medium	Reputation High	Reputation NA	Offer	TransactionUtility
Supplier 1	0.5	0.5	0	0	discrete	6
Supplier 2	0	0.5	0.5	0	sufficient	7.5

4.2 Price Variation Based on Past Experience

As explained in Sec. 3.4, each agent of our system changes the selling price considering its experience in past collaborations. As all the agents will change their prices, the conditions to organize the supply chain are dynamic and, therefore, it is possible to observe how the strategies adopted by the agents affect the price of the product provided by the whole supply chain and by each sub-chain.

To these purposes, we run 100 simulations where the agents negotiate to organize a supply chain and then automatically modify their offers in order to increase the chance of being selected for the supply chain creation and to maximize their profit. In the experiments, we set γ to 0 and τ to 2, with a temporal window of 3 and 2 respectively. To show how the agents modify their own behaviors, we focus on negotiations about the same product.

Let us consider the competition between suppliers S_4 and S_5 for the product P_2. Assuming that the two agents have the same reputation distribution, then the customer agent will select the supplier based on the best offer. Across time, to join the chain, agents S_4 and S_5 decrease their selling prices in order to be more competitive (see Fig. 5). The same scenario but with different reputation distributions for the agents is showed in Fig. 6. Now S_4 decreases its offered price faster than S_5 to reduce the reputation gap. When S_5 decreases its price too, the agent S_5 is preferred because of its reputation.

Let us consider the competition between suppliers S_2 and S_3 for the product P_1. We consider the case when S_3 always has a better reputation distribution. In this case the agents have a different minimum price but the lowest is not always selected because of the worst agent reputation(see Fig. 7). In all the plots in Figs. 5, 6 and 7, the markers at each run correspond to the agent that has been select to join the supply chain.For clarity, we used square and circle dots on the dashed and continue curves respectively. At each time, the dot indicates the agent winning the competition.

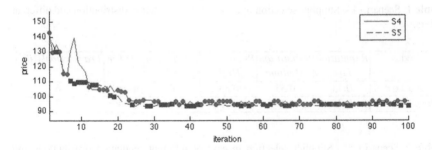

Fig. 5 The figure shows how the price of the product P_2 offered by S_4 and S_5 changes across time. The agents have the same reputation distributions and the supplier offering the lowest price is selected. Therefore, agents tends to decrease their prices to be more competitive.

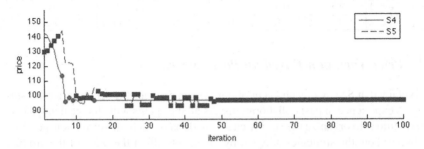

Fig. 6 The figure shows the plot of the product P_2 price offered by S_4 and S_5. Agent S5 has a better reputation distribution, and when a minimum tradeoff price is reached it is always selected from the higher level.

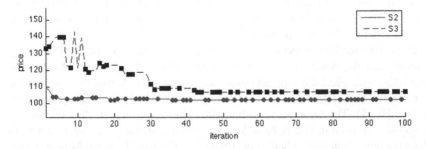

Fig. 7 The figure shows how the prices of the product P_1 offered by S_2 and S_3 changes across time. In this case, the agents have different reputation distributions; the supplier selection is done based on the transaction utility taking into account both the price and the reputation.

5 Conclusion and Future Works

In this paper we presented a decision support system for the automatic organization of a supply chain. In our formulation, a supply chain can be modeled hierarchically as a tree where each node represents a company providing a certain product/service to the higher level and buying products/services from the lower level. In practice, each sub-tree models a sub-chain. We employed agents for representing the companies involved in the supply chain organization and equipped each agent with a BDN they can adopt to filter information flowing in the customer/supplier network. In particular, the proposed BDN explicitly models the uncertainty of the information owned by the agent and related to the dynamic environment the agent works in. Moreover, our BDN formalizes the concept of reputation of the suppliers and permits to select those suppliers that would guarantee the best utility for the agent and the success for the whole supply chain.

We presented a simplified model that can be further improved through the collaboration between the knowledge engineer and a domain expert. The domain expert will identify the best decision criterion to consider in the supply chain management and their role. After a careful analysis a general model will be defined and customized by each company according to its management policies. Therefore in future works, more complex scenarios will be analyzed and more attention will be paid in modeling the decisions of each agent of the chain. In particular, we will extend the proposed BDN to model decisions about the possibility of joining the supply chain (based on potential risks the company would avoid or its resource availability) and the decisions about the supplier choice considering other variables such as time constraints and goodness of the agreement conditions or constraints between the suppliers at the same level of the chain.

Acknowledgments. This work has been supported by FRASI (Framework for Agent-based Semantic-aware Interoperability), an industrial research project funded by Italian Ministry of Education and Research.

References

1. Klein, M.R., Methlie, L.B.: Knowledge-Based Decision Support Systems: with Applications in Business, 2nd edn. John Wiley and Sons, Inc. (1995)
2. Chan, H.K., Chan, F.T.S.: Comparative study of adaptability and flexibility in distributed manufacturing supply chains. Decision Support Systems 48(2), 331–341 (2010), ISSN 0167-9236, doi:10.1016/j.dss.2009.09.001
3. Datta, P.P., Christopher, M.G.: Information sharing and coordination mechanisms for managing uncertainty in supply chains: a simulation study. International Journal of Production Research 49(3), 765–803 (2011)
4. Kumar, V., Srinivasan, S.: A Review of Supply Chain Management using Multi-Agent System. International Journal of Computer Science Issues 7(5) (September 2010)
5. Sycara, K.P.: Multiagent systems. AI Magazine 19(2), 79–92 (1998)
6. Wooldridge, M.: Intelligent agents. In: Gerhard, W. (ed.) Multiagent Systems: A Modern Approach to Distributed Artificial Intelligence, ch. 1, pages 2778. The MIT Press (1999)

7. Russell, S.J., Norvig, P.: Artificial Intelligence: A Modern Approach, 2nd edn. Pearson Education (2003)
8. A Multi-Agent Decision Support System for Dynamic Supply Chain Organization. In: Proceedings of the 5th International Workshop on New Challenges in Distributed Information Filtering and Retrieval (DART 2011), Palermo, Italy, September 17 (2011)
9. Jain, V., Wadhwa, S., Deshmukh, S.G.: Revisiting information systems to support a dynamic supply chain: issues and perspectives. Production Planning and Control: The Management of Operations 20(1), 17–29 (2009)
10. Sadeh, N.M., Hildum, D.W., Kjenstad, D.: Agent-Based E-Supply Chain Decision Support. Journal of Organizational Computing and Electronic Commerce 13(3 and 4), 225–241 (2003)
11. Moyaux, T., Chaib-Draa, B.: Supply Chain Management and Multiagent Systems: An Overview. In: Chaib-Draa, B., Mller, J.P. (eds.) Multiagent-Based Supply Chain Management, pp. 1–27 (2006)
12. Collins, J., Ketter, W., Sadeh, N.: Pushing the limits of rational agents: the Trading Agent Competition for Supply Chain Management. AI Magazine 31(2) (Summer 2010); Also available as Technical Report CMU-ISR-09-129
13. Zhang, Z., Tao, L.: Multi-agent Based Supply Chain Management with Dynamic Reconfiguration Capability. In: Proceedings of the 2008 IEEE/WIC/ACM International Conference on Web Intelligence and Intelligent Agent Technology (WI-IAT 2008), vol. 02, pp. 92–95. IEEE Computer Society, Washington, DC (2008), http://dx.doi.org/10.1109/WIIAT.2008.276, doi:10.1109/WIIAT.2008.276
14. Piramuthu, S.: Machine learning for dynamic multi-product supply chain formation. Expert Systems with Applications 29(4), 985–990 (2005) ISSN: 0957-4174, doi:10.1016/j.eswa.2005.07.004
15. Guneri, A.F., Yucel, A., Ayyildiz, G.: An integrated fuzzy-lp approach for a supplier selection problem in supply chain management. Expert Systems with Applications 36, 9223–9228 (2009)
16. Smith, R.G.: The contract net protocol: High-level communication and control in a distributed problem solver. IEEE Transactions on Computers 100(12), 1104–1113 (1980)
17. Hsieh, F.-S.: Analysis of contract net in multi-agent systems. Automatica 42(5), 733–740 (2006) ISSN 00051098
18. Wu, B., Cheng, T., Yang, S., Zhang, Z.: Price-based negotiation for task assignment in a distributed network manufacturing mode environment. The International Journal of Advanced Manufacturing Technology 21(2), 145–156 (2003)
19. Van Brussel, H., Wyns, J., Valckenaers, P., Bongaerts, L., Peeters, P.: Reference architecture for holonic manufacturing systems: PROSA. Computers in Industry 37(3), 255–274 (1998)
20. Alibhai, Z.: What is Contract Net Interaction Protocol? IRMS Lab. SFU (July 2003)
21. Lam, K.-C., Tao, R., La, M.C.-K.: A materialsupplier selection model for property developers using Fuzzy Principal Component Analysis. Automation in Construction 19, 608–618 (2010)
22. Chen, Y., Peng, Y.: An Extended Bayesian Belief Network Model of Multi-agent Systems for Supply Chain Managements. In: Truszkowski, W., Hinchey, M., Rouff, C.A. (eds.) WRAC 2002. LNCS (LNAI), vol. 2564, pp. 335–346. Springer, Heidelberg (2003)
23. JADE, http://jade.tilab.com/
24. GeNIe, http://genie.sis.pitt.edu/

Author Index